Directed Reading A

AF344471

Section: The Diversity of Cells

1. The smallest unit that can perform all the processes necessary for life is

a(n) _______________________________.

CELLS AND THE CELL THEORY

Match the correct description with the correct name. Write the letter in the space provided.

_______ **2.** He was the first person to describe cells.

_______ **3.** He discovered single-celled organisms including bacteria.

_______ **4.** He concluded that all plant parts were made of cells.

_______ **5.** He concluded that all animal tissues were made of cells.

_______ **6.** He concluded that all cells come from existing cells.

a. Schleiden

b. Virchow

c. Hooke

d. Leeuwenhoek

e. Schwann

CELL SIZE

_______ **7.** Why can a chicken egg grow so large?
   a. It is a single cell.
   b. It has a yolk and a shell.
   c. It does not have to take in nutrients.
   d. It grows faster than small cells.

_______ **8.** What limits most cells to a very small size?
   a. the surface area–to-volume ratio of the cell
   b. the thickness of the cell membrane
   c. the amount of cytoplasm in the cell
   d. the number of surrounding cells

_______ **9.** How would you calculate the surface area–to-volume ratio?
   a. Divide the volume by the surface area.
   b. Divide the total surface area of the cell by the cell's volume.
   c. Multiply the area of each side times the number of sides.
   d. Multiply the surface area times the volume.

Directed Reading A *continued*

10. What are the three parts of the cell theory?

____

____

____

11. What kind of cells have cell walls?

____

PARTS OF A CELL

Match the correct description with the correct term. Write the letter in the space provided.

________**12.** a protective layer that covers a cell's surface

________**13.** the fluid inside a cell

________**14.** a structure that performs a specific function in the cells

________**15.** the genetic material that carries information needed to make new cells or new organisms

________**16.** an organelle that contains DNA and has a role in growth, metabolism, and reproduction

a. DNA

b. cell membrane

c. nucleus

d. organelle

e. cytoplasm

TWO KINDS OF CELLS

17. What parts do all cells have in common?

____

18. What are the two basic kinds of cells?

____

PROKARYOTES: EUBACTERIA AND ARCHAEBACTERIA

19. What are prokaryotes?

____

Directed Reading A *continued*

20. What are the most common prokaryotes (and the smallest cells)?

____

21. What are ribosomes?

____

____

22. How do eubacteria and archaebacteria differ?

____

____

____

23. What are three types of archaebacteria?

____

____

____

EUKARYOTIC CELLS AND EUKARYOTES

______**24.** How do eukaryotes compare in size to prokaryotes?
  a. Eukaryotes have more cells.
  b. They are about the same size.
  c. Eukaryotes are about 10 times smaller.
  d. Eukaryotes are about 10 times larger.

______**25.** What does a eukaryote have that a prokaryote does not?
  a. one or more cells
  b. cells with a nucleus
  c. cells with DNA
  d. cells with membranes

______**26.** Which of these words describes humans?
  a. eukaryote
  b. prokaryote
  c. protist
  d. fungus

27. What does "multicellular" mean?

____

Directed Reading A

Section: Eukaryotic Cells

CELL WALL

1. What is the function of a cell wall?

____

2. What are the cell walls of plants and algae made of?

____

3. What are the cell walls of fungi made of?

____

CELL MEMBRANE

4. What is a cell membrane?

____

____

____

5. What are three types of compounds contained in the cell membrane?

____

6. What two substances control the movement of materials into and out of the cell?

____

____

CYTOSKELETON

_______ **7.** A web of proteins in the cytoplasm is known as the

    a. phospholipid.   **c.** cell membrane.

    b. cytoskeleton.   **d.** organelle.

8. What are the two functions of the cytoskeleton?

____

____

____

Directed Reading A *continued*

NUCLEUS

_______ **9.** What is the genetic material contained inside a cell's nucleus?
    a. protein   **c.** DNA
    b. lipids   **d.** nucleolis

_______ **10.** The function of proteins in a cell is to
    a. control chemical reactions.   **c.** cover the nucleus.
    b. store genetic information.   **d.** copy messages from DNA.

_______ **11.** What is the nucleolus?
    a. the opposite of the nucleus
    b. another name for DNA
    c. a network of fibers in the cytoplasm
    d. a dark area of the nucleus that stores materials and begins
        to make ribosomes

RIBOSOMES

12. Organelles that make proteins are called _______________________.

13. Proteins are made of _______________________.

ENDOPLASMIC RETICULUM

14. A system of folded membranes in which proteins, lipids and other materials

are made is the _______________________.

15. Two forms of endoplasmic reticulum are _______________________

and _______________________.

MITOCHONDRIA

_______ **16.** What function does a mitochondrion perform?
    a. It breaks down sugar to produce energy.
    b. It makes proteins.
    c. It breaks down toxic materials.
    d. It stores material used to make ribosomes.

17. The site of cellular respiration is the _______________________.

18. Energy produced in mitochondria is stored in a substance called

_______________________.

Directed Reading A *continued*

CHLOROPLASTS

_______**19.** Chloroplasts are organelles that are found in the cells of
    a. animals.
    b. plants and algae.
    c. mitochondria.
    d. all eukaryotic cells.

_______**20.** Which process happens inside a chloroplast?
    a. production of ATP
    b. production of DNA
    c. photosynthesis
    d. formation of animal cells

_______**21.** Chloroplasts are green because they contain
    a. sugar.
    b. proteins.
    c. chlorophyll.
    d. DNA.

GOLGI COMPLEX

_______**22.** The function of the Golgi complex is to
    a. produce sugar and water.
    b. package and deliver proteins.
    c. produce oxygen.
    d. trap energy from the sun.

CELL COMPARTMENTS

23. A small sac that surrounds material to be moved into or out of a cell

is a(n) _______________________ .

CELLULAR DIGESTION

24. What is a lysosome?

____

____

25. What is the function of lysosomes?

____

____

26. What function do vacuoles perform in plant and fungal cells?

____

____

Directed Reading A

Section: The Organization of Living Things
THE BENEFITS OF BEING MULTICELLULAR

1. How do multicellular organisms grow?

____

2. What are three benefits of being multicellular?

____

____

____

CELLS WORKING TOGETHER

3. What is a tissue?

____

____

4. What are four basic types of tissues in animals?

____

____

____

5. What are three basic types of tissues in plants?

____

____

____

TISSUES WORKING TOGETHER

6. A structure that is made up of two or more tissues working together is

called a(n) ________________________.

ORGANS WORKING TOGETHER

7. A group of organs working together to perform a particular function is

called a(n) ________________________.

Directed Reading A *continued*

8. What are examples of plant organs?

____

ORGANISMS

_______ **9.** Anything that can perform life processes is
   a. a cell.
   b. an organ system.
   c. a tissue.
   d. an organism.

_______ **10.** The term for any organism with only one cell is
   a. protist.
   b. unicellular.
   c. specialized.
   d. bacteria.

_______ **11.** Which of these is the lowest level of organization?
   a. cells
   b. tissues
   c. organs
   d. systems

_______ **12.** Which of these is the highest level of organization?
   a. cells
   b. tissues
   c. organs
   d. organ systems

STRUCTURE AND FUNCTION

13. The arrangement of parts in an organism is the ____________________________.

14. The job the part does within the organism is the ____________________________.

15. The millions of tiny air sacs in the lungs are called ____________________________.

Directed Reading B

Section: The Diversity of Cells

Circle the letter of the best answer for each question.

1. Which phrase contains the most important fact about cells?

    a. discovered with microscopes

    b. are basic units of life

    c. are too small to see

    d. discovered by accident

CELLS AND CELL THEORY

Read the words in the box. Read the sentences. Fill in each blank with the word that best completes the sentence.

animals	cells	microscope	plants

2. Robert Hooke was the first person to describe

    _______________________________.

3. Hooke built a(n) _______________________________ and used it to

    look at cells.

4. Hooke spent most of his time looking at the cells

    of _______________________________.

5. Hooke's microscope could not see the cells

    of _______________________________.

Directed Reading B *continued*

Finding Cells in Other Organisms

<u>Circle the letter</u> of the best answer for each question.

6. Where did Leeuwenhoek find what he called *animalcules*?

a. in animal blood

b. in bread dough

c. in cells

d. in pond scum

The Cell Theory

7. Which of these is not a part of the cell theory?

a. Most cells are too small to be seen without a microscope.

b. All organisms are made of one or more cells

c. The cell is the basic unit of all living things.

d. All cells come from existing cells.

CELL SIZE
A Few Large Cells

8. Why can a chicken egg grow so large?

a. It is a single cell.

b. It has a yolk and a shell.

c. It does not have to take in food.

d. It grows faster than small cells.

Many Small Cells

9. What limits most cells to a very small size?

a. the surface area–to-volume ratio

b. the size of the nucleus

c. the amount of fluid in the cell

d. the hardness of the cell wall

Directed Reading B *continued*

PARTS OF A CELL

The Cell Membrane and Cytoplasm

Read the words in the box. Read the sentences. <u>Fill in each blank</u> with the word or phrase that best completes the sentence.

cells	cytoplasm	cell membrane

10. The layer that protects a cell from its environment is

the _______________________.

11. The fluid inside a cell is called _______________________.

12. The cell membrane and cytoplasm are two parts of

all _______________________.

Organelles

<u>Circle the letter</u> of the best answer for each question.

13. Which sentence is true about most organelles?

a. They float outside the cell.

b. They have specific functions in the cell.

c. They live in green algae.

d. They are always the same.

Genetic Material

14. What is the organelle which contains the cell's DNA called?

a. membrane

b. nucleus

c. cell wall

d. cytoplasm

TWO KINDS OF CELLS

Read the words in the box. Read the sentences. <u>Fill in each blank</u> with the word or phrase that best completes the sentence.

cells	eukaroytic	prokaryotic

15. The two groups of _______________________ are eukaryotic and prokaryotic.

16. Cells that are _______________________ have a nucleus.

17. Cells that are _______________________ do not have a nucleus.

PROKARYOTES: EUBACTERIA AND ARCHAEBACTERIA

Eubacteria

<u>Circle the letter</u> of the best answer for each question.

18. What is the common name for eubacteria?

   a. prokaryote

   b. ribosome

   c. bacteria

   d. flagellum

Archaebacteria

19. What is one way in which archaebacteria differ from eubacteria?

   a. Archaebacteria lack of a nucleus.

   b. Archaebacteria have a cell membrane.

   c. Archaebacteria are single-celled.

   d. Archaebacterial ribosomes are different.

Directed Reading B *continued*

Circle the letter of the best answer for each question.

20. Which group includes extremophiles?

   a. eubacteria

   b. archaebacteria

   c. methane gases

   d. eukaryotes

EUKARYOTIC CELLS AND EUKARYOTES

21. What does a eukaryote have that a prokaryote does not?

   a. one or more cells

   b. cells with a nucleus

   c. cells with DNA

   d. cells with membranes

22. Which of these words describes you and other humans?

   a. eukaryote

   b. prokaryote

   c. protist

   d. fungus

Skills Worksheet

Directed Reading B

Section: Eukaryotic Cells
CELL WALL
<u>Circle the letter</u> of the best answer for each question.

1. What is the purpose of a cell wall?

   a. to make a plant droop

   b. to support the cell

   c. to carry DNA

   d. to digest cellulose

CELL MEMBRANE

2. What is the purpose of a cell membrane?

   a. to make lipids

   b. to make phospholipids

   c. to protect the cell

   d. to support the cell wall

3. What does having two layers allow the cell membrane to do?

   a. make lipids and phospholipids

   b. support the cell wall

   c. make proteins

   d. pass nutrients and wastes through

CYTOSKELETON

4. What is the cytoskeleton made of?

   a. cells

   b. lipids

   c. membranes

   d. proteins

Circle the letter of the best answer for each question.

5. What is the cytoskeleton's job in the cell?

   a. help keep the cell's shape

   b. process proteins

   c. store water

   d. produce energy

NUCLEUS

6. What is the genetic material inside a cell's nucleus?

   a. protein

   b. lipids

   c. DNA

   d. nucleolis

7. What is the function of proteins in a cell?

   a. to control chemical reactions

   b. to store genetic information

   c. to cover the nucleus

   d. to copy messages from DNA

8. What is an amino acid?

   a. part of the cell membrane

   b. another term for DNA

   c. a dangerous chemical

   d. a molecule used to make proteins

Directed Reading B continued

RIBOSOMES

<u>Circle the letter</u> of the best answer for each question.

9. What do all ribosomes do?

  a. make proteins

  b. float in the cytoplasm

  c. attach themselves to membranes

  d. make organelles

ENDOPLASMIC RETICULUM

10. What does the endoplasmic reticulum look like?

  a. oval, with pores

  b. small and round

  c. long, with many folds

  d. a bubble full of liquid

11. Which phrase tells the function of the endoplasmic reticulum?

  a. internal delivery system

  b. protein factory

  c. DNA storage

  d. web of proteins

MITOCHONDRIA

Circle the letter of the best answer for each question.

12. What are the peanut-shaped organelles that break down sugar?

    a. Golgi complex

    b. cell membranes

    c. ribosomes

    d. mitochondria

CHLOROPLASTS

13. Which process happens inside a chloroplast?

    a. making ATP

    b. making DNA

    c. photosynthesis

    d. formation of animal cells

GOLGI COMPLEX

14. What long, folded cell part serves to package and distribute proteins?

    a. Golgi complex

    b. cell membrane

    c. ribosome

    d. cytoplasm

CELL COMPARTMENTS

15. Why do vesicles move around the cytoplasm?

    a. to make new proteins

    b. to move material around

    c. to support the cell membrane

    d. to form the Golgi complex

| Directed Reading B *continued*

CELLULAR DIGESTION

Circle the letter of the best answer for each question.

16. What do lysosomes do?

    a. make new proteins

    b. move material around

    c. get rid of waste and digest food

    d. create vesicles

Vacuoles

17. What is a function of some vacuoles?

    a. to make proteins   **c.** to make sugar

    b. to store water   **d.** to harden the cell

ORGANELLES AND THEIR FUNCTIONS

The Cell Membrane and Cytoplasm

Read the words in the box. Read the sentences. Fill in each blank with the word or phrase that best completes the sentence.

chloroplasts	endoplasmic reticulum	lysosomes
mitochondria	nucleus	

18. The cell part that contains most of the DNA is

    the ________________________.

19. The part that makes lipids and breaks down drugs is

    the ________________________.

20. Cell parts that break down molecules to make ATP are

    called ________________________.

21. Plant cell parts that use the sun to make food are

    called ________________________.

22. Vesicles that break down food particles and cellular wastes

    are called ________________________.

Directed Reading B

Section: The Organization of Living Things
THE BENEFITS OF BEING MULTICELLULAR

Circle the letter of the best answer for each question.

1. Which of these is a benefit of being a large organism?

    a. cell specialization

    b. larger cells

    c. smaller cells

    d. shorter lifespan

CELLS WORKING TOGETHER

2. What is a *tissue* made of?

    a. cells that work together

    b. larger than normal cells

    c. cells with longer lives

    d. cardiac muscle

3. What organisms have nerve, muscle, connective, and protective tissues?

    a. animals   **c.** fungi

    b. plants   **d.** cardiac muscles

TISSUES WORKING TOGETHER

4. What is made up of two or more tissues working together?

    a. a cell   **c.** an organ

    b. a connective tissue   **d.** a group of specialized cells

5. Which of these is a plant organ?

    a. heart   **c.** blood

    b. leaf   **d.** transport tissue

Directed Reading B *continued*

ORGANS WORKING TOGETHER

<u>Circle the letter</u> of the best answer for each question.

6. What do you call a group of organs that work together to perform a particular function?

a. connective organs

c. an organ system

b. an organism

d. a human being

ORGANISMS

7. What is the term for anything that can perform a life process?

a. a cell

b. an organ system

c. an organization

d. an organism

8. Which of these is the lowest level of organization?

a. cells

c. organs

b. tissues

d. organ systems

STRUCTURE AND FUNCTION

9. What is the word for the arrangement of parts in an organism?

a. function

b. structure

c. cell

d. shape

10. What is a part's structure related to?

a. its function

b. its material

c. support from its cells

d. its ability to get rid of wastes

Vocabulary and Section Summary

The Diversity of Cells

VOCABULARY

In your own words, write a definition of the following terms in the space provided.

1. cell

____

____

2. cell membrane

____

____

3. organelle

____

____

4. nucleus

____

____

5. prokaryote

____

____

6. eukaryote

____

____

SECTION SUMMARY

Read the following section summary.

- Cells were not discovered until microscopes were invented in the 1600s.

- Cell theory states that all organisms are made of cells, the cell is the basic unit of all living things, and all cells come from other cells.

- All cells have a cell membrane, cytoplasm, and DNA.

- Most cells are too small to be seen with the naked eye. A cell's surface area–to-volume ratio limits the size of a cell.

- The two basic kinds of cells are prokaryotic cells and eukaryotic cells. Eukaryotic cells have a nucleus and membrane-bound organelles. Prokaryotic cells do not.

- Prokaryotes are classified as archaebacteria and eubacteria.

- Archaebacterial cell walls and ribosomes are different from the cell walls and ribosomes of other organisms.

- Eukaryotes can be single-celled or multicellular.

Vocabulary and Section Summary

Eukaryotic Cells

VOCABULARY

In your own words, write a definition of the following terms in the space provided.

1. cell wall

____

____

2. ribosome

____

____

3. endoplasmic reticulum

____

____

4. mitochondrion

____

____

5. Golgi complex

____

____

6. vesicle

____

____

7. lysosome

____

____

Vocabulary and Section Summary *continued*

SECTION SUMMARY

Read the following section summary.

- Eukaryotic cells have organelles that perform functions that help cells remain alive.
- All cells have a cell membrane. Some cells have a cell wall. Some cells have a cytoskeleton.
- The nucleus of a eukaryotic cell contains the cell's genetic material, DNA.
- Ribosomes are the organelles that make proteins. Ribosomes are not covered by a membrane.
- The endoplasmic reticulum (ER) and the Golgi complex make and process proteins before the proteins are transported to other parts of the cell or out of the cell.
- Mitochondria and chloroplasts are energy-producing organelles.
- Lysosomes are organelles responsible for digestion within a cell. In plant cells, organelles called vacuoles store cell materials and sometimes act like large lysosomes.

Vocabulary and Section Summary

The Organization of Living Things
VOCABULARY

In your own words, write a definition of the following terms in the space provided.

1. tissue

____

____

2. organ

____

____

3. organ system

____

____

4. organism

____

____

5. structure

____

____

6. function

____

____

SECTION SUMMARY

Read the following section summary.

- Advantages of being multicellular are larger size, longer life, and cell specialization.

- Four levels of organization are cell, tissue, organ, and organ system.

- A tissue is a group of cells working together to do a specific job. An organ is two or more tissues working together to do a specific job. An organ system is two or more organs working together to do a specific job.

- In organisms, a part's structure and function are related.

Section Review

The Diversity of Cells

USING KEY TERMS

1. In your own words, write a definition for the term *organelle*.

____

____

2. Use the following terms in the same sentence: *prokaryotic, nucleus,* and *eukaryotic.*

____

____

____

UNDERSTANDING KEY IDEAS

______ **3.** Cell size is limited by the
  a. thickness of the cell wall.
  b. size of the cell's nucleus.
  c. cell's surface area–to-volume ratio.
  d. amount of cytoplasm in the cell.

4. What are the three parts of the cell theory?

____

____

____

5. Name three structures that every cell has.

____

____

____

6. Give two ways in which archaebacteria are different from bacteria.

____

____

| Section Review *continued*

CRITICAL THINKING

7. Applying Concepts You have discovered a new single-celled organism. It has a cell wall, ribosomes, and long, circular DNA. Is it a eukaryote or a prokaryote cell? Explain.

____

____

8. Identifying Relationships One of your students brings you a cell about the size of the period at the end of this sentence. It is a single cell, but it also forms chains. What characteristics would this cell have if the organism is a eukaryote? If it is a prokaryote? What would you look for first?

____

____

INTERPRETING GRAPHICS

The picture below shows a particular organism. Use the picture to answer the questions that follow.

9. What type of organism does the picture represent? How do you know?

____

____

10. Which structure helps the organism move?

____

____

11. What part of the organism does the letter *A* represent?

____

____

Section Review

Eukaryotic Cells

USING KEY TERMS

1. In your own words, write a definition for each of the following terms: *ribosome, lysosome,* and *cell wall.*

UNDERSTANDING KEY IDEAS

_______ **2.** Which of the following are found mainly in animal cells?
   a. mitochondria
   b. lysosomes
   c. ribosomes
   d. Golgi complexes

3. What is the function of a Golgi complex? What is the function of the endoplasmic reticulum?

CRITICAL THINKING

4. Making Comparisons Describe three ways in which plant cells differ from animal cells.

5. Applying Concepts Every cell needs ribosomes. Explain why.

Section Review *continued*

6. Predicting Consequences A certain virus attacks the mitochondria in cells. What would happen to a cell if all of its mitochondria were destroyed?

____

____

7. Expressing Opinions Do you think that having chloroplasts gives plant cells an advantage over animal cells? Support your opinion.

____

____

____

____

INTERPRETING GRAPHICS

Use the diagram below to answer the questions that follow.

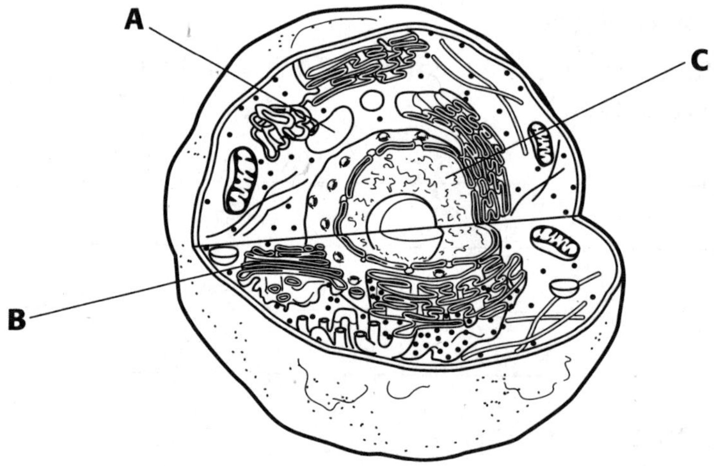

8. Is this a diagram of a plant cell or an animal cell? Explain how you know.

____

____

9. What organelle does the letter *B* refer to?

____

____

Section Review

The Organization of Living Things

USING KEY TERMS

1. Use each of the following terms in a separate sentence: *tissue, organ,* and *function.*

____

____

____

UNDERSTANDING KEY IDEAS

______ 2. What are the four levels of organization in living things?
   a. cell, multicellular, organ, organ system
   b. single cell, multicellular, tissue, organ
   c. larger size, longer life, specialized cells, organs
   d. cell, tissue, organ, organ system

MATH SKILLS

3. One multicellular organism is a cube. Each of its sides is 3 cm long. Each of its cells is 1 cm^3. How many cells does it have? If each side doubles in length, how many cells will it then have? Show your work below.

CRITICAL THINKING

4. **Applying Concepts** Explain the relationship between structure and function. Use alveoli as an example. Be sure to include more than one level of organization.

____

____

____

____

Section Review *continued*

5. **Making Inferences** Why can multicellular organisms be more complex than single-cell organisms? Use the three advantages of being multicellular to help explain your answer.

____

____

____

____

Chapter Review

USING KEY TERMS

Complete each of the following sentences by choosing the correct term from the word bank.

cell	organ	cell membrane
prokaryote	organelles	eukaryote
cell wall	tissue	structure
function		

1. A(n) ________________________ is the most basic unit of all living things.

2. The job that an organ does is the ________________________ of that organ.

3. Ribosomes and mitochondria are types of ________________________.

4. A(n) ________________________ is an organism whose cells have a nucleus.

5. A group of cells working together to perform a specific function is

a(n) ________________________.

6. Only plant cells have a(n) ________________________.

UNDERSTANDING KEY IDEAS

Multiple Choice

_______ **7.** Which of the following best describes an organ?
   a. a group of cells that work together to perform a specific job
   b. a group of tissues that belong to different systems
   c. a group of tissues that work together to perform a specific job
   d. a body structure, such as the heart or lungs

_______ **8.** The benefits of being multicellular include
   a. small size, long life, and cell specialization.
   b. generalized cells, longer life, and ability to prey on small animals.
   c. larger size, more enemies, and specialized cells.
   d. longer life, larger size, and specialized cells.

_______ **9.** In eukaryotic cells, which organelle contains the DNA?
   a. nucleus   **c.** smooth ER
   b. Golgi complex   **d.** vacuole

_______ **10.** Which of the following statements is part of the cell theory?
   a. All cells suddenly appear by themselves.
   b. All cells come from other cells.
   c. All organisms are multicellular.
   d. All cells have identical parts.

| Chapter Review *continued*

______**11.** The surface area–to-volume ratio of a cell limits
  a. the number of organelles that the cell has.
  b. the size of the cell.
  c. where the cell lives.
  d. the types of nutrients that a cell needs.

______**12.** Two types of organisms whose cells do not have a nucleus are
  a. prokaryotes and eukaryotes.
  b. plants and animals.
  c. eubacteria and archaebacteria.
  d. single-celled and multicellular organisms.

Short Answer

13. Explain why most cells are small.

____

____

14. Describe the four levels of organization in living things.

____

____

15. What is the difference between the structure of an organ and the function of the organ?

____

____

16. Name two functions of a cell membrane.

____

____

17. What are the structure and function of the cytoskeleton in a cell?

____

____

Chapter Review *continued*

CRITICAL THINKING

18. Concept Mapping Use the following terms to create a concept map: *cells, organisms, Golgi complex, organ systems, organs, nucleus, organelle* and *tissues.*

| Chapter Review *continued*

19. Making Comparisons Compare and contrast the functions of the endoplasmic reticulum and the Golgi complex.

____

____

____

20. Identifying Relationships Explain how the structure and function of an organism's parts are related. Give an example.

____

____

____

21. Evaluating Hypotheses One of your classmates states a hypothesis that all organisms must have organ systems. Is your classmate's hypothesis valid? Explain your answer.

____

____

____

22. Predicting Consequences What would happen if all of the ribosomes in your cells disappeared?

____

____

____

23. Expressing Opinions Scientists think that millions of years ago the surface of the Earth was very hot and that the atmosphere contained a lot of methane. In your opinion, which type of organism, a eubacterium or an archaebacterium, is the older form of life? Explain your reasoning.

____

____

____

____

INTERPRETING GRAPHICS

Use the diagram below to answer the questions that follow.

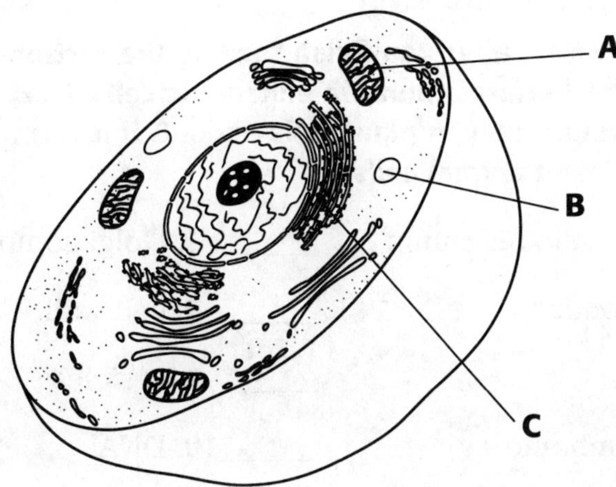

24. What is the name of the structure identified by the letter *A*?

____

25. Which letter identifies the structure that digests food particles and foreign invaders?

____

26. Which letter identifies the structure that makes proteins, lipids, and other materials and that contains tubes and passageways that enable substances to move to different places in the cell?

____

Reinforcement

Building a Eukaryotic Cell

**Complete this worksheet after you finish reading the section "Eukaryotic Cells."
Below is a list of the features found in eukaryotic cells. Next to each feature, write
P if it is a feature found only in plant cells and a B if it is a feature that can be
found in both plant and animal cells.**

_______ **1.** endoplasmic reticulum	_______ **7.** Golgi complex
_______ **2.** mitochondria	_______ **8.** cell wall
_______ **3.** nucleus	_______ **9.** vesicles
_______ **4.** cell membrane	_______**10.** DNA
_______ **5.** cytoplasm	_______**11.** nucleolus
_______ **6.** ribosomes	_______**12.** chloroplasts

**In the space provided, label the structures of the eukaryotic cell drawn below.
Include only the structures that you labeled B.**

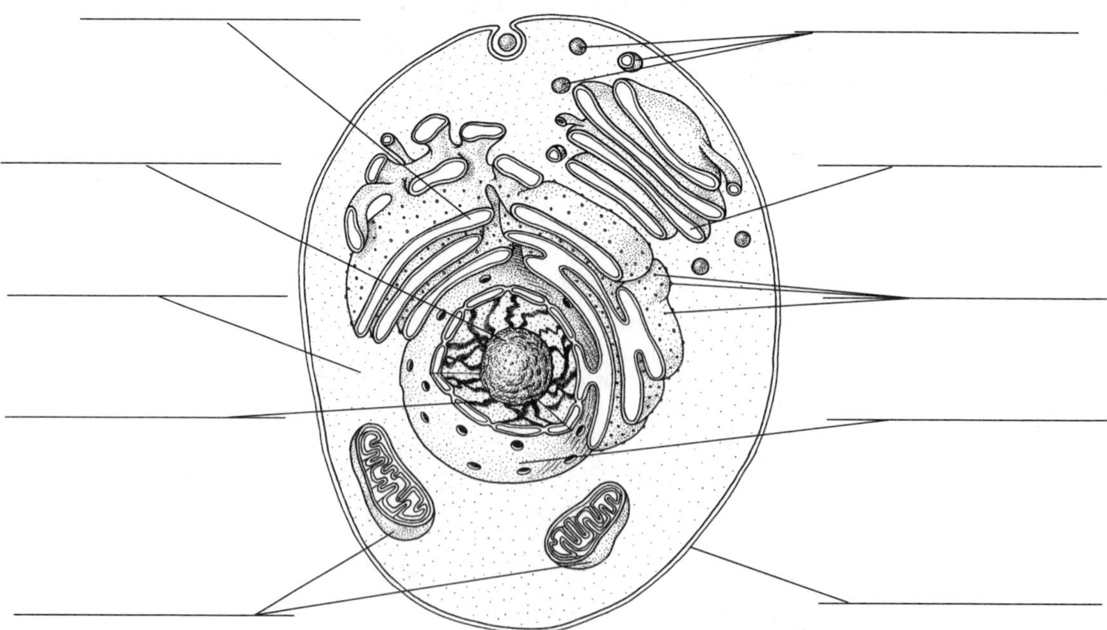

Critical Thinking

Cellular Construction

Mr. Dan Plumbob
Patche & Splinte, Architects

Dear Mr. Plumbob:

We are pleased to inform you that we have chosen your firm to design our new Megalopolis factory. As you know, I have always been fascinated by the biology of cells. Therefore, I have decided to model our new factory after the cell of an animal.

Your job is to design each part of the factory so that it represents a cellular structure. I have faith in your firm's ability to complete this project.

Sincerely,

____
(your signature here)

SEEING RELATIONSHIPS

1. In the chart below, each part of a cell is compared to a part of a factory. In the right column of the chart, write in the function performed by each pair of components. The first one is done for you.

Cell components	Factory components	Functions
cell membrane	perimeter fence	protects and controls access
mitochondria	energy generators	
nucleus	director's office	
lysosome	waste management	
endoplasmic reticulum	materials delivery system	
Golgi complex	packaging department	

Critical Thinking *continued*

MAKING COMPARISONS

2. How might the architectural plans for building the Megalopolis factory be compared to the DNA in cells?

____

____

____

____

____

3. What advantages might there be to having a factory that functioned more like a plant cell than like an animal cell? Explain your answer.

____

____

____

____

____

____

DESIGN YOUR OWN

4. Design and label a diagram for a Megalopolis factory that operates like an animal cell. Include all the factory components listed in the chart. Label each factory component with its cellular counterpart.

Section Quiz

Section: The Diversity of Cells

Match the correct definition or description with the correct term. Write the letter in the space provided.

_______ **1.** tiny, round organelles made of protein and other material

_______ **2.** the fluid inside a cell

_______ **3.** the reason that most cells are limited to a very small size

_______ **4.** a protective layer that covers the cell's surface and acts as a barrier

_______ **5.** small bodies in a cell's cytoplasm that are specialized to perform specific functions

_______ **6.** in a eukaryotic cell, an organism that contains the cell's DNA and that has a role in growth, metabolism, and reproduction

_______ **7.** an organism that consists of a single cell that does not have a nucleus or membrane-bound organelles

_______ **8.** prokaryotes that are the smallest cells and that have ribosomes

_______ **9.** prokaryotes that include extremophiles, organisms that live in extreme conditions

_______ **10.** an organism made up of cells that have a nucleus enclosed by a membrane as well as membrane-bound organelles

a. archaebacteria

b. cell membrane

c. eubacteria

d. eukaryote

e. nucleus

f. organelle

g. prokaryote

h. ribosomes

i. surface area–to-volume ratio

j. cytoplasm

Section Quiz

Section: Eukaryotic Cells

Match the correct definition with the correct term. Write the letter in the space provided.

_______ **1.** a rigid structure that gives support to a cell

_______ **2.** a barrier that encloses and protects the cell

_______ **3.** a web of proteins in the cytoplasm that keeps a cell's membrane from collapsing

_______ **4.** a large organelle that produces and stores the cell's DNA

_______ **5.** organelles that make proteins

_______ **6.** a system of folded membranes that functions as the internal delivery system of a cell

_______ **7.** an organelle that functions as the main power source of a cell, breaking down sugar to produce energy

_______ **8.** organelles in which photosynthesis takes place

_______ **9.** the organelle that packages and distributes proteins

_______ **10.** organelles that contain digestive enzymes

a. cell membrane

b. cell wall

c. chloroplasts

d. cytoskeleton

e. endoplasmic reticulum

f. Golgi complex

g. lysosomes

h. mitochondrion

i. nucleus

j. ribosomes

Section Quiz

Section: The Organization of Living Things

Write the letter of the correct answer in the space provided.

______ **1.** Humans like you are
- **a.** machines.
- **b.** systems.
- **c.** organisms.
- **d.** protists.

______ **2.** One benefit of being a large organism is that you have
- **a.** larger cells.
- **b.** fewer predators.
- **c.** simpler functions.
- **d.** only one kind of cell.

______ **3.** The life span of a multicellular organism is
- **a.** only as long as the life of one cell.
- **b.** shorter than that of a single-celled organism.
- **c.** not limited to the life of a single cell.
- **d.** the same in every cell.

______ **4.** A group of cells with the same function make up
- **a.** an organism.
- **b.** an organ system.
- **c.** a tissue.
- **d.** a structure.

______ **5.** In what kind of tissue does photosynthesis take place?
- **a.** nerve
- **b.** muscle
- **c.** transport
- **d.** ground

______ **6.** An organ consists of
- **a.** two or more tissues.
- **b.** a group of cells.
- **c.** two or more systems.
- **d.** nerves and muscles.

______ **7.** An organ system has
- **a.** one kind of tissue.
- **b.** only one function.
- **c.** two or more organs.
- **d.** one main kind of cell.

______ **8.** Even simple multicellular organisms can have
- **a.** organs.
- **b.** specialized cells.
- **c.** systems.
- **d.** colonies.

______ **9.** The highest level of organization is the
- **a.** cell.
- **b.** tissue.
- **c.** organ.
- **d.** system.

______ **10.** The functions of an organism's parts are related to those parts'
- **a.** structures.
- **b.** systems.
- **c.** blood cells.
- **d.** alveoli.

Chapter Test A

Cells: The Basic Units of Life

MULTIPLE CHOICE

Write the letter of the correct answer in the space provided.

______ **1.** What is smallest unit that can perform all the processes necessary for life?
   a. cell
   b. nucleus
   c. organelle
   d. protist

______ **2.** Robert Hooke and Anton van Leeuwenhoek not only helped discover cells but also
   a. discovered that cells came from existing cells.
   b. helped develop the microscope.
   c. concluded that all living things have cells.
   d. discovered mushrooms and fungi.

______ **3.** Leeuwenhoek called the single-celled organisms that he found in pond scum "animalcules." Today we know them as
   a. animals.
   b. plant life.
   c. fungi.
   d. protists.

______ **4.** Scientist Matthias Schleiden contributed to the cell theory by concluding that
   a. the cells of plants and animals were the same.
   b. all plant parts were made of cells.
   c. the cells of plants were different from those of animals.
   d. all animal tissues were made of cells.

______ **5.** Which of the following statements is not part of the cell theory?
   a. Animals and plants share the same kinds of cells.
   b. All organisms are made up of one or more cells.
   c. The cell is the basic unit of all living things.
   d. All cells come from existing cells.

______ **6.** Most cells are a very small size because
   a. they don't have hard shells like eggs.
   b. their volume does not increase.
   c. their volume is limited by how large their surface area is.
   d. their surface area–to-volume ratio is too small.

Chapter Test A *continued*

MATCHING

Match the correct definition with the correct term. Write the letter in the space provided.

______ **7.** the part of the cell that keeps the cytoplasm inside and controls materials going in and out of the cell

______ **8.** structures that are usually surrounded by membranes and which perform specific functions within the cell

______ **9.** an organelle that contains the cell's deoxyribonucleic acid

______ **10.** a single-celled organism that has no nucleus or membrane-bound organelles

______ **11.** the smallest and most common form of prokaryotes, containing DNA, ribosomes, and a flagellum

______ **12.** prokaryotes that include such types as heat-loving, salt-loving, and methane-making

______ **13.** organisms made up of cells that have a nucleus and membrane-bound organelles

______ **14.** word that describes most organisms that you can see with your naked eye

a. archaebacteria

b. cell membrane

c. eubacteria

d. eukaryotes

e. multicellular

f. nucleus

g. organelles

h. prokaryote

MULTIPLE CHOICE

Write the letter of the correct answer in the space provided.

______ **15.** What cell part supports the cell and might be made of cellulose or chitin?
  a. cell membrane
  b. cell wall
  c. ribosome
  d. nucleus

______ **16.** What part of the cell forms a barrier between the cell and its environment?
  a. cell membrane
  b. cell wall
  c. ribosome
  d. cholesterol

Chapter Test A continued

_______ **17.** What part of the cell keeps the cell membrane from collapsing?
    a. cell wall
    b. cytoplasm
    c. cytoskeleton
    d. nucleus

_______ **18.** A cell's nucleus contains DNA, which carries genetic material with
    a. ribosomes.
    b. the cytoskeleton.
    c. the endoplasmic reticulum.
    d. instructions for how to make protein.

_______ **19.** Ribosomes, the organelles that make proteins, are found on the membranes of the
    a. cell wall.
    b. endoplasmic reticulum.
    c. mitochondria.
    d. vacuoles.

_______ **20.** What part of the cell acts as the cell's delivery system?
    a. nucleus
    b. nucleolis
    c. mitochondrion
    d. endoplasmic reticulum

_______ **21.** Energy released by a cell's mitochondrion is stored in
    a. ATP.
    b. DNA.
    c. the ER.
    d. RNA.

_______ **22.** What cell parts carry materials between organelles such as the ER and the Golgi complex?
    a. ribosomes
    b. lysosomes
    c. vesicles
    d. vacuoles

Chapter Test A *continued*

______**23.** Larger size, longer life, and specialization are three advantages to
being a
 a. eukaryote.
 b. prokaryote.
 c. unicellular organism.
 d. multicellular organism.

______**24.** Which of the following is true of each of the four levels of organization
of living things?
 a. Each contains larger cells than the level below it.
 b. Each is more complex than the level below it.
 c. Each performs the same functions as the level below it.
 d. Each is more specialized than the level below it.

______**25.** The function of a part of an organism is related to
 a. its arrangement of cells.
 b. the shape of its parts.
 c. the structure of that part.
 d. its appearance under a microscope.

Chapter Test B

Cells: The Basic Units of Life

USING KEY TERMS

Use the terms from the following list to complete the sentences below. Each term may be used only once. Some terms may not be used.

cell membrane	nucleus	mitochondria
lysome	ribosome	organs
organelle	Golgi complex	prokaryotic

1. Various tissues that work together to perform a specific job

constitute a(n) ______________________.

2. The role of the cell's ______________________ is to release energy that can

be used to power various cellular processes.

3. DNA, the genetic material in cells, is located in a eukaryotic

cell's ______________________.

4. Cells that have no membrane-covered organelles are

______________________.

5. A part of the Golgi complex can pinch off and form

a(n) ______________________, which distributes materials to other parts of

the cell.

UNDERSTANDING KEY IDEAS

Write the letter of the correct answer in the space provided.

_______ **6.** Which statement is NOT part of the cell theory?
  a. All organisms are made of one or more cells.
  b. Animal and plant cells contain the same organelles.
  c. The cell is the basic unit of living things.
  d. All cells originate from other cells.

_______ **7.** A cell's volume grows faster than its surface area, so if a cell gets
  too large
  a. its surface area–to-volume ratio will decrease.
  b. the cell membrane and cell walls will break down.
  c. its outer surface will harden like an eggshell does.
  d. it will not be able to take in enough nutrients or get rid of wastes.

Chapter Test B *continued*

_______ **8.** A large vesicle that aids in digestion within plant cells the way lysosomes do is called
- **a.** an enzyme.
- **b.** a vacuole.
- **c.** a mitochondrion.
- **d.** a nucleolus.

_______ **9.** Most of a cell's ATP is made and stored in the inner membrane of the
- **a.** Golgi complex.
- **b.** nucleus.
- **c.** endoplasmic reticulum.
- **d.** mitochondrion.

_______ **10.** Specialization in cells makes tissues, organs, and systems
- **a.** grow large in size.
- **b.** produce larger cells.
- **c.** work more efficiently.
- **d.** stay healthy.

11. What four elements do all cells have in common?

____

____

12. What are three elements that plant cells have and animal cells do not?

____

____

13. List three roles played by proteins within a cell.

____

____

____

14. Identify two functions of the Golgi complex and describe how it performs those functions.

____

____

____

15. List four levels of organization of living things.

____

____

____

CRITICAL THINKING

16. Compare the levels of organization among eukaryotes with the types of organization found among prokaryotes.

____

____

____

____

17. Why does being single-celled give a species of extremophiles a greater chance of survival? To answer the question, review your answer to question 16 and imagine a multicellular species in the same environment.

____

____

____

____

18. Explain why the cells in an embryo will grow no larger than a certain size before they divide.

____

____

____

____

____

19. Explain how the structure of alveoli makes their function possible.

____

____

____

____

____

| Chapter Test B *continued*

COMPLETING A CONCEPT MAP

20. Use the following terms to complete the concept map below:

DNA   Golgi complex   Mitochondria
proteins   ribosomes

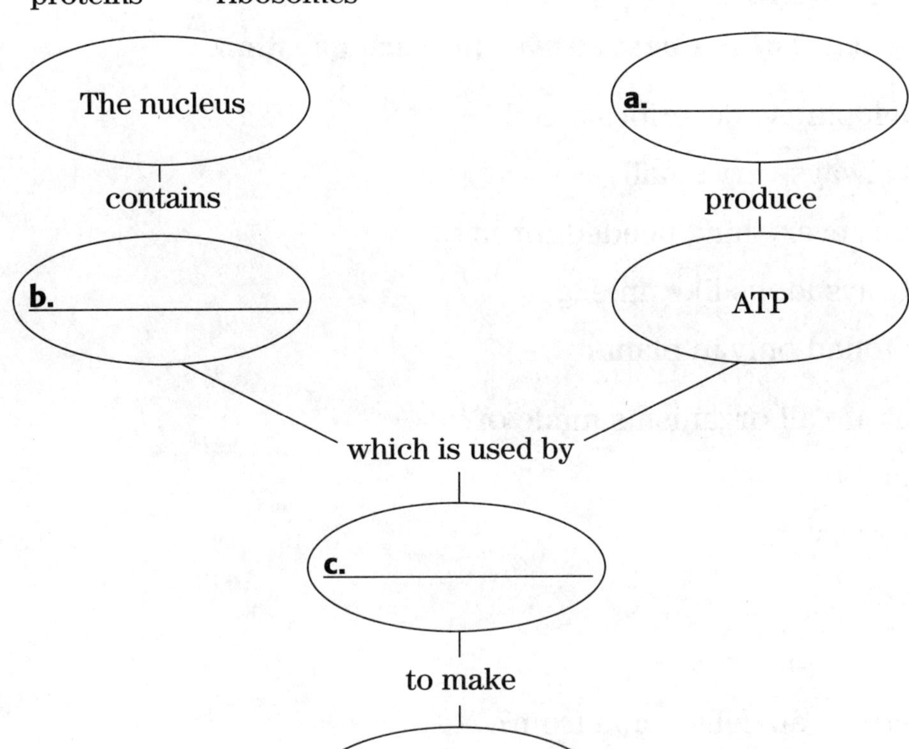

Chapter Test C

Cells: The Basic Units of Life
MULTIPLE CHOICE
Circle the letter of the best answer for each question.

1. Which phrase describes a cell?

  a. is always very small

  b. does everything needed for life

  c. always looks like an egg

  d. is found only in plants

2. What are all organisms made of?

  a. plants

  b. protists

  c. cells

  d. eggs

3. Where do all cells come from?

  a. animals

  b. ponds

  c. cells

  d. eggs

4. What keeps the size of most cells very small?

  a. their hard shells

  b. the surface area–to-volume ratio

  c. food and wastes

  d. their thin surfaces

5. What protects the inside of a cell from the outside world?

  a. cytoplasm   **c.** cell membrane

  b. nucleus   **d.** DNA

| Chapter Test C *continued*

Circle the letter of the best answer for each question.

6. How are archaebacteria different from eubacteria?

   a. Archaebacteria have different ribosomes.

   b. Archaebacteria have only one cell.

   c. Archaebacteria have cell membranes.

   d. Archaebacteria have RNA, not DNA.

7. What is cytoplasm?

   a. the nucleus of a cell

   b. the fluid inside a cell

   c. the genetic material in a cell

   d. the proteins in a cell

8. Where does photosynthesis take place in a cell?

   a. in the nucleus

   b. in the mitochondria

   c. in the chloraplasts

   d. in the ribosomes

9. What does the Golgi complex do in a cell?

   a. It packages and distributes proteins.

   b. It is the power source of the cell.

   c. It makes sugar and oxygen.

   d. It makes proteins.

10. What is the job of the lysosomes?

   a. They store water.

   b. They digest food particles

   c. They make new cells.

   d. They package proteins.

| Chapter Test C *continued*

MATCHING

Read the description. Then, <u>draw a line</u> from the dot next to each description to the matching word.

11. a cell with a nucleus ●   **a.** DNA

12. a cell without a nucleus ●   **b.** eukaryote

13. genetic material in cells ●   **c.** nucleus

14. where DNA is stored ●   **d.** prokaryote

---

15. stiff surfaces that support cells ●   **a.** cell walls

16. organelle that makes proteins ●   **b.** endoplastmic reticulum

17. a cell's delivery system ●   **c.** ribosome

FILL-IN-THE-BLANK

Read the words in the box. Read the sentences. <u>Fill in each blank</u> with the word or phrase that best completes the sentence.

cell	multicellular	organ
structure	system	tissue

18. The lowest level of organization is

the _______________________________.

19. Cells that are like each other and do the same job form

a(n) _______________________________.

20. A structure made of two or more tissues working together is

called a(n) _______________________________.

21. A group of organs that work together form an

organ _______________________________.

22. Larger size, longer life, and more-specialized cells are

characteristics of _______________________________ organisms.

23. How a part of an organism works is related to how it is built,

or its _______________________________.

Performance-Based Assessment

OBJECTIVES

In this activity you will use everyday materials to build models of plant, animal, and prokaryotic cells.

KNOW THE SCORE!

As you work through the activity, keep in mind that you will be earning a grade for the following:

- how you work with materials (10%)
- how well you build your cell models (50%)
- how well you complete the analysis (40%)

MATERIALS

- 3 resealable sandwich bags
- 3 clear plastic jars just large enough to hold a sandwich bag
- light corn syrup or gelatin
- small items such as beads, beans, pipe cleaners, grapes, pasta spirals, thread, yarn, etc.

PROCEDURES

1. Using the materials provided, build a model of a plant cell. The plant cell should have nine of the structures listed in the box.

2. Build a model of an animal cell. The animal cell should have seven of the structures listed in the box.

3. Build a model of a prokaryotic cell. The prokaryotic cell should have four of the structures listed in the box.

CELLULAR STRUCTURES

Some structures can be used more than once.

• cell membrane	• lysosomes
• cell wall	• mitochondria
• chloroplast	• nucleus
• endoplasmic reticulum	• ribosomes
• Golgi complex	• vacuole
• circular DNA	

Performance-Based Assessment *continued*

ANALYSIS

4. In the charts on the next page, indicate which material you used to represent each of the cell structures.

5. How are prokaryotic cells similar to eukarytic cells?

____

____

____

6. How are plant cells different from animal cells?

____

____

____

▌Performance-Based Assessment *continued*

Plant Cell

Part of cell	Material used in model

Animal Cell

Part of cell	Material used in model

Prokaryotic Cell

Part of cell	Material used in model

Standardized Test Preparation

READING

Read each of the passages below. Then, answer the questions that follow each passage.

Passage 1 Exploring caves can be dangerous but can also lead to interesting discoveries. For example, deep in the darkness of Cueva de Villa Luz, a cave in Mexico, are slippery formations called *snottites*. They were named snottites because they look just like a two-year-old's runny nose. If you use an electron microscope to look at them, you see that snottites are bacteria; thick, sticky fluids; and small amounts of minerals produced by the bacteria. As tiny as they are, these bacteria can build up snottite structures that may eventually turn into rock. Formations in other caves look like hardened snottites. The bacteria in snottites are acidophiles. Acidophiles live in environments that are highly acidic. Snottite bacteria produce sulfuric acid and live in an environment that is similar to the inside of a car battery.

_______ **1.** Which statement best describes snottites?
   A Snottites are bacteria that live in car batteries.
   B Snottites are rock formations found in caves.
   C Snottites were named for a cave in Mexico.
   D Snottites are made of bacteria, sticky fluids, and minerals.

_______ **2.** Based on this passage, which conclusion about snottites is most likely to be correct?
   F Snottites are found in caves everywhere.
   G Snottite bacteria do not need sunlight.
   H You could grow snottites in a greenhouse.
   I Snottites create other bacteria in caves.

_______ **3.** What is the main idea of this passage?
   A Acidophiles are unusual organisms.
   B Snottites are strange formations.
   C Exploring caves is dangerous.
   D Snottites are large, slippery bacteria.

Standardized Test Preparation *continued*

Passage 2 The world's smallest mammal may be a bat about the size of a jelly bean. The scientific name for this tiny animal, which was unknown until 1974, is *Craseonycteris thonglongyai*. It is so small that it is sometimes called the *bumblebee bat*. Another name for this animal is the *hog-nosed bat*. Hog-nosed bats were given their name because one of their distinctive features is a piglike muzzle. Hog-nosed bats differ from other bats in another way: they do not have a tail. But, like other bats, hog-nosed bats do eat insects that they catch in mid-air. Scientists think that the bats eat small insects that live on the leaves at the tops of trees. Hog-nosed bats live deep in limestone caves and have been found in only one country, Thailand.

______ **1.** According to the passage, which statement about hog-nosed bats is most accurate?
  A They are the world's smallest animal.
  B They are about the size of a bumblebee.
  C They eat leaves at the tops of trees.
  D They live in hives near caves in Thailand.

______ **2.** Which of the following statements describes distinctive features of hog-nosed bats?
  F The bats are very small and eat leaves.
  G The bats live in caves and have a tail.
  H The bats live in Thailand and are birds.
  I The bats have a piglike muzzle and no tail.

______ **3.** From the information in this passage, which conclusion is most likely to be correct?
  A Hog-nosed bats are similar to other bats.
  B Hog-nosed bats are probably rare.
  C Hog-nosed bats can sting like a bumblebee.
  D Hog-nosed bats probably eat fruit.

Standardized Test Preparation *continued*

INTERPRETING GRAPHICS

The diagrams below show two kinds of cells. Use these cell diagrams to answer the questions that follow.

Cell 1

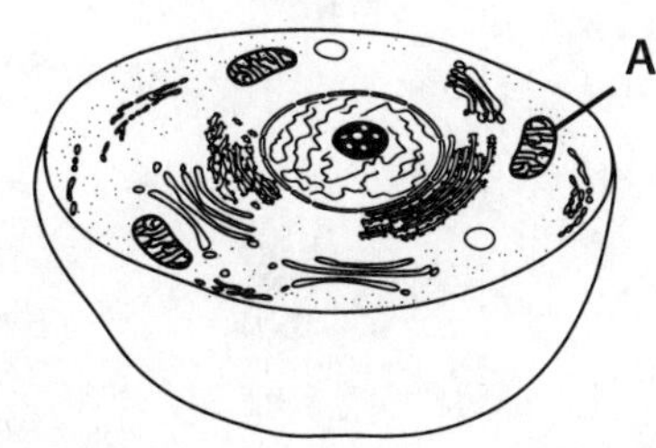

Cell 2

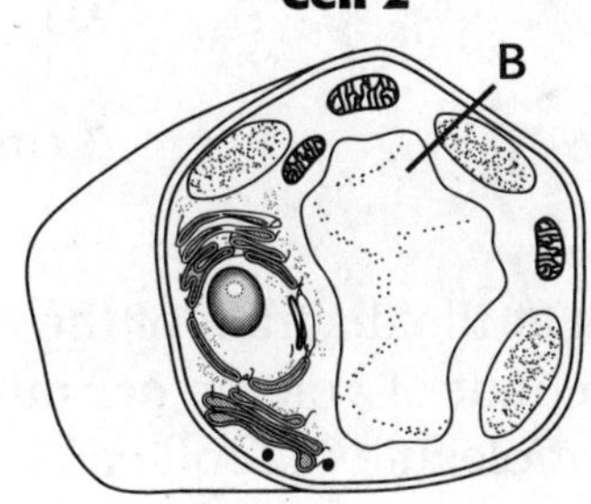

_______ **1.** What is the name of the organelle labeled A in Cell 1?
   A endoplasmic reticulum
   B mitochondrion
   C vacuole
   D nucleus

_______ **2.** What type of cell is Cell 1?
   F a bacterial cell
   G a plant cell
   H an animal cell
   I a prokaryotic cell

_______ **3.** What is the name and function of the organelle labeled B in Cell 2?
   A The organelle is a vacuole, and it stores water and other materials.
   B The organelle is the nucleus, and it contains the DNA.
   C The organelle is the cell wall, and it gives shape to the cell.
   D The organelle is a ribosome, where proteins are put together.

_______ **4.** What type of cell is Cell 2? How do you know?
   F prokaryotic; because it does not have a nucleus
   G eukaryotic; because it does not have a nucleus
   H prokaryotic; because it has a nucleus
   I eukaryotic; because it has a nucleus

Standardized Test Preparation *continued*

MATH

Read each question below, and choose the best answer.

_______ **1.** What is the surface area–to-volume ratio of the rectangular solid shown in the diagram below?

   A 0.5:1
   B 2:1
   C 36:1
   D 72:1

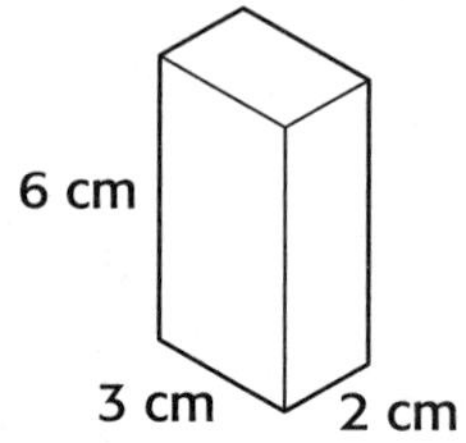

_______ **2.** Look at the diagram of the cell below. Three molecules of food per cubic unit of volume per minute are required for the cell to survive. One molecule of food can enter through each square unit of surface area per minute. What will happen to this cell?

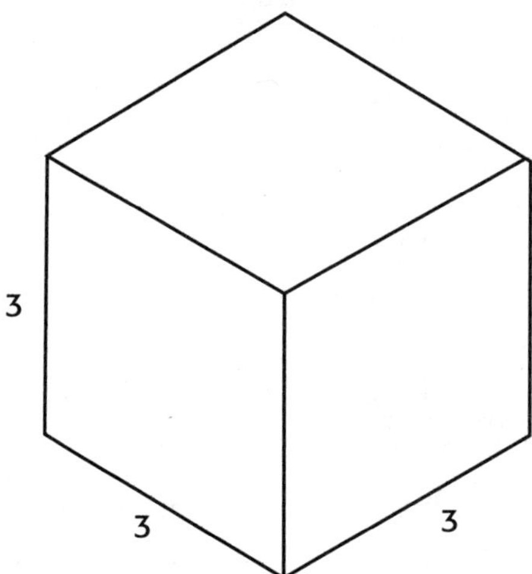

   F The cell is too small, and it will starve.
   G The cell is too large, and it will starve.
   H The cell is at a size that will allow it to survive.
   I There is not enough information to determine the answer.

Model-Making Lab

Elephant-Sized Amoebas?

An amoeba is a single-celled organism. Like most cells, amoebas are microscopic. Why can't amoebas grow as large as elephants? If an amoeba grew to the size of a quarter, the amoeba would starve to death. To understand how this can be true, build a model of a cell and see for yourself.

OBJECTIVES

Explore why a single-celled organism cannot grow to the size of an elephant.

Create a model of a cell to illustrate the concept of surface area–to-volume ratio.

MATERIALS

- calculator (optional)
- cubic cell patterns
- heavy paper or poster board
- sand, fine
- scale or balance
- scissors
- tape, transparent

SAFETY INFORMATION 

PROCEDURE

1. Use heavy paper or poster board to make four cube-shaped cell models from the patterns supplied by your teacher. Cut out each cell model, fold the sides to make a cube, and tape the tabs on the sides. The smallest cell model has sides that are each one unit long. The next larger cell has sides of two units. The next cell has sides of three units, and the largest cell has sides of four units. These paper models represent the cell membrane, the part of a cell's exterior through which food and wastes pass.

Data Table for Measurements					Key to Formula Symbols
Length of side	**Area of one side** (A = S × S)	**Total surface area of cube cell** (TA = S × S × 6)	**Volume of cube cell** (V = S × S × S)	**Mass of filled cube cell**	S = the length of one side
1 unit	1 unit^2	6 unit^2	1 unit^2		A = area
2 unit					6 = number of sides
3 unit					V = volume
4 unit					TA = total area

2. Fill in the data table above. Use each formula to calculate the data about your cell models. Record your calculations in the table. Calculations for the smallest cell have been done for you.

Elephant-Sized Amoebas? *continued*

3. Carefully fill each model with fine sand until the sand is level with the top edge of the model. Find the mass of the filled models by using a scale or a balance. What does the sand in your model represent?

____

4. Record the mass of each filled cell model in your Data Table for Measurements. (Always remember to use the appropriate mass unit.)

____

ANALYZE THE RESULTS

1. Constructing Tables Make a data table like the one shown below.

Data Table for Ratios		
Length of side	Ratio total surface area to volume	Ratio of total surface area to mass
1 unit		
2 unit		
3 unit		
4 unit		

2. Organizing Data Use the data from your Data Table for Measurements to find the ratios for each of your cell models. For each of the cell models, fill in the Data Table for Ratios.

Elephant-Sized Amoebas? *continued*

DRAW CONCLUSIONS

3. Interpreting Information As a cell grows larger, does the ratio of total surface area to volume increase, decrease, or stay the same?

____

4. Interpreting Information As a cell grows larger, does the total surface area–to-mass ratio increase, decrease, or stay the same?

____

5. Drawing Conclusions Which is better able to supply food to all the cytoplasm of the cell: the cell membrane of a small cell or the cell membrane of a large cell? Explain your answer.

____

____

____

____

____

6. Evaluating Data In the experiment, which is better able to feed all of the cytoplasm of the cell: the cell membrane of a cell that has high mass or the cell membrane of a cell that has low mass? You may explain your answer in a verbal presentation to the class, or you may choose to write a report and illustrate it with drawings of your models.

____

____

____

____

____

(Quick Lab)

Bacteria in Your Lunch?

MATERIALS

- cotton swab
- coverslip, plastic
- microscope
- microscope slide, plastic
- water
- yogurt with active culture

PROCEDURE

Most of the time, you don't want bacteria in your food. Many bacteria make toxins that will make you sick. However, some foods—such as yogurt—are supposed to have bacteria in them! The bacteria in these foods are not dangerous.

In yogurt, masses of rod-shaped bacteria feed on the sugar (lactose) in milk. The bacteria convert the sugar into lactic acid. Lactic acid causes milk to thicken. This thickened milk makes yogurt.

1. Using a **cotton swab**, put a **small dot of yogurt** on a **microscope slide**.

2. Add a **drop of water**. Use the cotton swab to stir.

3. Add a **coverslip**.

4. Use a **microscope** to examine the slide. Draw what you observe.

Name _________________________________ Class _______________ Date ____________

Cells Alive!

DATASHEET FOR LABBOOK

You have probably used a microscope to look at single-celled organisms. They can be found in pond water. In the following exercise, you will look at *Protococcus*—algae that form a greenish stain on tree trunks, wooden fences, flowerpots, and buildings.

MATERIALS

- eyedropper
- microscope
- microscope slide and coverslip
- *Protococcus* (or other algae)
- water

SAFETY INFORMATION 

PROCEDURE

1. Locate some *Protococcus*. Scrape a small sample into a container. Bring the sample to the classroom, and make a wet mount of it as directed by your teacher. If you can't find *Protococcus* outdoors, look for algae on the glass in an aquarium. Such algae may not be *Protococcus*, but it will be a very good substitute.

2. Set the microscope on low power to examine the algae. On a separate sheet of paper, draw the cells that you see.

3. Switch to high power to examine a single cell. Draw the cell.

4. You will probably notice that each cell contains several chloroplasts. Label a chloroplast on your drawing. What is the function of the chloroplast?

____

____

5. Another structure that should be clearly visible in all the algae cells is the nucleus. Find the nucleus in one of your cells, and label it on your drawing. What is the function of the nucleus?

____

____

6. What does the cytoplasm look like? Describe any movement you see inside the cells.

____

____

Cells Alive! *continued*

ANALYZE THE RESULTS

1. Are *Protococcus* single-celled organisms or multicellular organisms?

____

____

2. How are *Protococcus* different from amoebas?

____

____

____

Vocabulary Activity

Cellular Crosswords

After you finish reading the chapter, use the clues below to complete the crossword puzzle on the next page.

ACROSS

3. the fluid inside a cell

7. the world's smallest cells

10. the chemical control center of a cell

11. organelle containing digestive enzymes

12. kind of cell that does not have a nucleus

13. organelle that packages and transport materials out of the cell

17. describes an organism that exists as a group of cells

20. A single _____ has everything necessary to carry out life's activities.

23. scientist who first described cells

24. energy-converting organelle found in plant and algae cells

25. anything that can live independently

26. groups of organs working together to perform particular jobs in the body

27. a structure performing a specific function within a cell

DOWN

1. The cells of plants and algae have a hard _____ _____ made of cellulose.

2. organelles that break down sugar to produce energy.

4. a combination of two or more tissues working together to perform a specific job in the body

5. organelles that make proteins

6. a group of similar cells that perform a common function

8. cells that have a nucleus and membrane-bound organelles

9. sacs that contain materials in a eukaryotic cell

14. barrier between the inside of a cell and its environment

15. dark area inside the nucleus that stores materials that will be used to make ribosomes

16. scientific description of all living things in terms of cells

18. the cell's delivery system (abbr.)

19. substance that stores energy released by mitochondria

21. a large vesicle that stores enzymes or liquids

22. the cell's hereditary material

Vocabulary Activity *continued*

Name _________________________________ Class _________________ Date _____________

SciLinks Activity

EUKARYOTIC CELLS

Go to www.scilinks.org. To find links related to eukaryotic cells, type in the keyword HSM0541. Then use the links to answer the following questions about cells that have a nucleus and membrane-bound organelles.

Go to www.scilinks.org

Topic: Eukaryotic Cells
SciLinks code: HSM0541

1. What is the term for the process of converting oxygen and nutrients into ATP?

____

____

2. What are microtubules, and where in the cell are they found?

____

____

____

3. Approximately how much older are prokaryotes than eukaryotes?

____

____

____

4. How many vacuoles are there in a typical plant cell?

____

____

____

5. How many individual molecules make up human DNA?

____

____

____

Performance-Based Assessment

Teacher Notes and Answer Key

PURPOSE

Students will review the structures of animal, plant, and prokaryotic cells by building models of them.

Rebecca Ferguson
Northridge Middle School
North Richland Hills, Texas

TIME REQUIRED

- The activity is designed for one 45-minute class period.
- Students will need 35 minutes to make and observe the model and 10 minutes to answer the analysis question.

RATING

Easy ◄——1——2——3——4——► Hard

Teacher Prep–3
Student Set-Up–2
Concept Level–3
Clean Up–2

ADVANCE PREPARATION

- Gather various small materials to represent organelles within model cells. Equip each student activity station with the necessary materials. Each group will need enough corn syrup or gelatin to fill all three jars.
- Have moist paper towels or rags available for cleaning up during model making.

SAFETY INFORMATION

Do not allow students to taste the materials. Clean up floor spills immediately to prevent slipping.

TEACHING STRATEGIES

- This activity works best in groups of 2–3 students.
- You may wish to administer this assessment as an open book exercise. Have students report to you when they have finished their models.

Performance-Based Assessment *continued*

Evaluation Strategies

Use the following rubric to help evaluate student performance.

Rubric for Assessment

Possible points	Quality and clarity of descriptions of the animal in its habitat (10 points possible)
10–5	Successfully completes activity; uses materials in an appropriate manner
4–1	Fails to take activity seriously; does not use materials in an appropriate manner; demonstrates unsafe behavior
	Construction of model cells (50 points possible)
50–40	Demonstrates superior knowledge of cellular structure; completes all three models; organelles well chosen; organelles represented in reasonable quantities
39–20	Demonstrates knowledge of cellular structure; completes at least two models; all organelles represented but not well chosen; organelle quantities incorrect
19–1	Demonstrates lack of knowledge of cellular structure; completes fewer than two models; some organelles missing or wrongly placed; organelle quantities incorrect
	Completion of analysis (40 points possible)
40–30	Accurately completes charts and questions; demonstrates exceptional understanding of cellular structures
29–1	Does not complete charts and/or questions; demonstrates lack of understanding or effort.

Name _____________________________ Class _______________ Date _____________

Performance-Based Assessment

OBJECTIVES

In this activity you will use everyday materials to build models of plant, animal, and prokaryotic cells.

KNOW THE SCORE!

As you work through the activity, keep in mind that you will be earning a grade for the following:

- how you work with materials (10%)
- how well you build your cell models (50%)
- how well you complete the analysis (40%)

MATERIALS

- 3 resealable sandwich bags
- 3 clear plastic jars just large enough to hold a sandwich bag
- light corn syrup or gelatin
- small items such as beads, beans, pipe cleaners, grapes, pasta spirals, thread, yarn, etc.

PROCEDURES

1. Using the materials provided, build a model of a plant cell. The plant cell should have nine of the structures listed in the box.
2. Build a model of an animal cell. The animal cell should have seven of the structures listed in the box.
3. Build a model of a prokaryotic cell. The prokaryotic cell should have four of the structures listed in the box.

CELLULAR STRUCTURES

Some structures can be used more than once.

• cell membrane	• lysosomes
• cell wall	• mitochondria
• chloroplast	• nucleus
• endoplasmic reticulum	• ribosomes
• Golgi complex	• vacuole
• circular DNA	

Name _______________________________ Class _______________ Date _______________

Performance-Based Assessment *continued*

ANALYSIS

4. In the charts on the next page, indicate which material you used to represent each of the cell structures.

5. How are prokaryotic cells similar to eukarytic cells?

Sample answer: Both types of cells are basic units of life that have a cell membrane, cytoplasm, DNA, and ribosomes.

6. How are plant cells different from animal cells?

Sample answer: Plant cells are capable of producing their own food and have different organelles, such as chloroplasts and cell walls, while animal cells do not.

Name _______________________________ Class _______________ Date _______________

Performance-Based Assessment *continued*

Plant Cell

Part of cell	Material used in model
nucleus	Sample answer: grape
ribosomes	Sample answer: red beads
cell membrane	Sample answer: plastic sandwich bag
mitochondria	Sample answer: peanuts or kidney beans
Golgi complex	Sample answer: spiral pasta
vacuole	Sample answer: bath oil bead, jelly bean
chloroplast	Sample answer: lima bean
cell wall	Sample answer: plastic container
endoplasmic reticulum	Sample answer: ribbon

Animal Cell

Part of cell	Material used in model
nucleus	Sample answer: grape
ribosomes	Sample answer: red beads
cell membrane	Sample answer: plastic sandwich bag
mitochondria	Sample answer: peanuts or kidney beans
Golgi complex	Sample answer: spiral pasta
endoplasmic reticulum	Sample answer: ribbon
lysomes	Sample answer: dried peas

Prokaryotic Cell

Part of cell	Material used in model
circular DNA	Sample answer: blue thread
ribosomes	Sample answer: red beads
cell membrane	Sample answer: plastic sandwich bag
cell wall	Sample answer: plastic container

Model-Making Lab)   **DATASHEET FOR CHAPTER LAB**

Elephant-Sized Amoebas?

Teacher Notes and Answer Key

TIME REQUIRED

Two 45-minute class periods

RATING

Easy ←—— 1 —— 2 —— 3 —— 4 —→ Hard

Teacher Prep–2
Student Set-Up–2
Concept Level–3
Clean Up–1

SAFETY CAUTION

Remind students to review all safety cautions and icons before beginning this lab
activity.

PREPARATION NOTES

Some students may find it difficult to work with a nonspecific unit of measurement. If so, the cube models easily convert to centimeters. You may want to add
some small items, such as peas, beans, popcorn, or peppercorns, to the sand to
represent organelles floating in the cytoplasm. Some students may need to review
what a ratio is and how ratios are used.

CELL MODEL TEMPLATE

Prepare four patterns for students to use to make their cubes.
Make one cube 1 unit wide, one cube 2 units wide, one cube 3 units wide,
and one cube 4 units wide. The unit can be the size of your choosing.

Name _______________________________ Class _______________ Date __________

  DATASHEET FOR CHAPTER LAB

Elephant-Sized Amoebas?

An amoeba is a single-celled organism. Like most cells, amoebas are microscopic. Why can't amoebas grow as large as elephants? If an amoeba grew to the size of a quarter, the amoeba would starve to death. To understand how this can be true, build a model of a cell and see for yourself.

OBJECTIVES

Explore why a single-celled organism cannot grow to the size of an elephant.

Create a model of a cell to illustrate the concept of surface area–to-volume ratio.

MATERIALS

- calculator (optional)
- cubic cell patterns
- heavy paper or poster board
- sand, fine
- scale or balance
- scissors
- tape, transparent

SAFETY INFORMATION 

PROCEDURE

1. Use heavy paper or poster board to make four cube-shaped cell models from the patterns supplied by your teacher. Cut out each cell model, fold the sides to make a cube, and tape the tabs on the sides. The smallest cell model has sides that are each one unit long. The next larger cell has sides of two units. The next cell has sides of three units, and the largest cell has sides of four units. These paper models represent the cell membrane, the part of a cell's exterior through which food and wastes pass.

Data Table for Measurements					Key to Formula Symbols
Length of side	**Area of one side** $(A = S \times S)$	**Total surface area of cube cell** $(TA = S \times S \times 6)$	**Volume of cube cell** $(V = S \times S \times S)$	**Mass of filled cube cell**	S = the length of one side
1 unit	1 unit^2	6 unit^2	1 unit^2		A = area
2 unit					6 = number of sides
3 unit					V = volume
4 unit					TA = total area

2. Fill in the data table above. Use each formula to calculate the data about your cell models. Record your calculations in the table. Calculations for the smallest cell have been done for you.

Name _______________________________ Class _________________ Date ____________

Elephant-Sized Amoebas? *continued*

3. Carefully fill each model with fine sand until the sand is level with the top edge of the model. Find the mass of the filled models by using a scale or a balance. What does the sand in your model represent?

The sand represents cytoplasm.

4. Record the mass of each filled cell model in your Data Table for Measurements. (Always remember to use the appropriate mass unit.)

Masses will vary.

ANALYZE THE RESULTS

1. Constructing Tables Make a data table like the one shown below.

Data Table for Ratios		
Length of side	Ratio total surface area to volume	Ratio of total surface area to mass
1 unit		
2 unit		
3 unit		
4 unit		

2. Organizing Data Use the data from your Data Table for Measurements to find the ratios for each of your cell models. For each of the cell models, fill in the Data Table for Ratios.

Name _________________________________ Class _______________ Date _______________

Elephant-Sized Amoebas? *continued*

DRAW CONCLUSIONS

3. Interpreting Information As a cell grows larger, does the ratio of total surface area to volume increase, decrease, or stay the same?

decreases

4. Interpreting Information As a cell grows larger, does the total surface area–to-mass ratio increase, decrease, or stay the same?

decreases

5. Drawing Conclusions Which is better able to supply food to all the cytoplasm of the cell: the cell membrane of a small cell or the cell membrane of a large cell? Explain your answer.

The cell membrane of a small cell. A small cell has a higher surface area-to-

volume ratio than a large cell has, so more nutrients per cubic unit of vol-

ume can enter a small cell.

6. Evaluating Data In the experiment, which is better able to feed all of the cytoplasm of the cell: the cell membrane of a cell that has high mass or the cell membrane of a cell that has low mass? You may explain your answer in a verbal presentation to the class, or you may choose to write a report and illustrate it with drawings of your models.

the cell membrane of a cell with low mass

Name _______________________________ Class ________________ Date ____________

DATASHEET FOR QUICK LAB

Bacteria in Your Lunch?

MATERIALS

- cotton swab
- coverslip, plastic
- microscope
- microscope slide, plastic
- water
- yogurt with active culture

PROCEDURE

Most of the time, you don't want bacteria in your food. Many bacteria make toxins that will make you sick. However, some foods—such as yogurt—are supposed to have bacteria in them! The bacteria in these foods are not dangerous.

In yogurt, masses of rod-shaped bacteria feed on the sugar (lactose) in milk. The bacteria convert the sugar into lactic acid. Lactic acid causes milk to thicken. This thickened milk makes yogurt.

1. Using a **cotton swab**, put a **small dot of yogurt** on a **microscope slide**.

2. Add a **drop of water**. Use the cotton swab to stir.

3. Add a **coverslip**.

4. Use a **microscope** to examine the slide. Draw what you observe.

   Drawings should depict rodshaped bacteria.

Cells Alive!

DATASHEET FOR LABBOOK

Teacher Notes and Answer Key

Terry Rakes
Elmwood Junior High School
Rogers, Arkansas

TIME REQUIRED

One 45-minute class period

RATING

Teacher Prep–1

Student Set-Up–1

Concept Level–1

Clean Up–1

SAFETY INFORMATION

MATERIALS

The materials listed on the student page are enough for a group of 3–4 students. Be sure to keep the algae in a warm, damp place out of direct sunlight; a closed plastic bag with water sprayed into it is ideal.

Name _______________________________ Class _______________ Date _____________

DATASHEET FOR LABBOOK

Cells Alive!

You have probably used a microscope to look at single-celled organisms. They can be found in pond water. In the following exercise, you will look at *Protococcus*—algae that form a greenish stain on tree trunks, wooden fences, flowerpots, and buildings.

MATERIALS

- eyedropper
- microscope
- microscope slide and coverslip
- *Protococcus* (or other algae)
- water

SAFETY INFORMATION 

PROCEDURE

1. Locate some *Protococcus*. Scrape a small sample into a container. Bring the sample to the classroom, and make a wet mount of it as directed by your teacher. If you can't find *Protococcus* outdoors, look for algae on the glass in an aquarium. Such algae may not be *Protococcus*, but it will be a very good substitute.

2. Set the microscope on low power to examine the algae. On a separate sheet of paper, draw the cells that you see.

3. Switch to high power to examine a single cell. Draw the cell.

4. You will probably notice that each cell contains several chloroplasts. Label a chloroplast on your drawing. What is the function of the chloroplast?

   Chloroplasts are the parts of the cell that are responsible for

   photosynthesis.

5. Another structure that should be clearly visible in all the algae cells is the nucleus. Find the nucleus in one of your cells, and label it on your drawing. What is the function of the nucleus?

   The nucleus of a cell controls most of the activities that take place in that

   cell and contains the hereditary information.

6. What does the cytoplasm look like? Describe any movement you see inside the cells.

   The cytoplasm is a clear gel-like substance that fills the cell and surrounds

   the organelles.

Name _______________________________ Class _______________ Date _______________

Cells Alive! *continued*

ANALYZE THE RESULTS

1. Are *Protococcus* single-celled organisms or multicellular organisms?

   Protococcus is a genus composed of single-celled algae. cannot move about like amoebas can; unlike amoebas, they are green and photosynthesize.

2. How are *Protococcus* different from amoebas?

   Many answers are possible, but the following are most likely: Protococcus cannot move about like amoebas can; unlike amoebas, they are green and photosynthesize.

Answer Key

Directed Reading A

SECTION: THE DIVERSITY OF CELLS

1. cell
2. C
3. D
4. A
5. E
6. B
7. C
8. A
9. B
10. All organisms are made of one or more cells. The cell is the basic unit of all living things. All cells come from existing cells.
11. cell of plants and fungi
12. B
13. E
14. D
15. A
16. C
17. cell membranes, organelles, cytoplasm, and DNA
18. eukaryotic and prokaryotic
19. Prokaryotes are organisms that consist of a single cell that does not have a nucleus or membrane-bound organelles.
20. eubacteria, or bacteria
21. tiny, round organelles made of protein and other material
22. Archaebacterial ribosomes are different from eubacterial ribosomes
23. heat-loving, salt-loving, and methane-making
24. D
25. B
26. A
27. "many cells"

SECTION: EUKARYOTIC CELLS

1. to give support to a cell
2. cellulose and other materials
3. chitin or a chemical similar to chitin
4. a protective layer that encloses the cell and separates the cell's contents from the cell's environment.
5. lipids, phospholipids, and proteins
6. proteins and lipids
7. B
8. to keep the cell's membrane from collapsing and to help its organelles move
9. C
10. A
11. D
12. ribosomes
13. amino acids
14. endoplasmic reticulum or ER
15. smooth, rough
16. A
17. a mitochondria
18. ATP
19. B
20. C
21. C
22. B
23. a vesicle
24. a lysosome is a vesicle responsible for digestion inside a cell.
25. Lysosomes destroy worn-out or damaged organelles, get rid of waste materials, and protect the cell from foreign invaders.
26. Vacuoles are large organelles that act like lysosomes or store water and other materials.

SECTION: THE ORGANIZATION OF LIVING THINGS

1. by making more cells
2. larger size, longer life, and specialization
3. A tissue is a group of similar cells that perform a common function.
4. nerve, muscle, connective, protective
5. transport, protective, ground
6. organ
7. organ system
8. leaves, stems, roots
9. D
10. B
11. A
12. D
13. structure
14. function
15. alveoli

Directed Reading B

SECTION: THE DIVERSITY OF CELLS

1. B	**12.** cells
2. cells	**13.** B
3. microscope	**14.** B
4. plants	**15.** cells
5. animals	**16.** eukaryotic
6. D	**17.** prokaryotic
7. A	**18.** C
8. C	**19.** D
9. A	**20.** B
10. cell membrane	**21.** B
11. cytoplasm	**22.** A

SECTION: EUKARYOTIC CELLS

1. B
2. C
3. D
4. D
5. A
6. C
7. A
8. D
9. A
10. C
11. A
12. D
13. C
14. A
15. B
16. C
17. B
18. nucleus
19. endoplasmic reticulum
20. mitochondria
21. chloroplasts
22. lysosomes

SECTION: THE ORGANIZATION OF LIVING THINGS

1. A
2. A
3. A
4. C
5. B
6. C
7. D
8. A
9. B
10. A

Vocabulary and Section Summary

SECTION: THE DIVERSITY OF CELLS

1. cell: in biology, the smallest unit that can perform all life processes; cells are covered by a membrane and have DNA and cytoplasm.

2. cell membrane: a phospholipid layer that covers a cell's surface; acts as a barrier between the inside of a cell and the cell's environment.

3. organelle: one of the small bodies in a cell's cytoplasm that are specialized to perform a specific function

4. nucleus: in a eukaryotic cell, an organelle that contains the cell's DNA and that has a role in growth, metabolism, and reproduction

5. prokaryote: an organism that consists of a single cell that does not have a nucleus or membrane-bound organelles

6. eukaryote: an organism made up of cells that have a nucleus enclosed by a membrane; eukaryotes include animals, plants, and fungi but not archaea or bacteria.

SECTION: EUKARYOTIC CELLS

1. cell wall: a rigid structure that surrounds the cell membrane and provides support to the cell

2. ribosome: a cell organelle composed of RNA and protein; the site of protein synthesis

3. endoplasmic reticulum: a system of membranes that is found in a cell's cytoplasm and that assists in the production, processing, and transport of proteins and in the production of lipids

4. mitochondrion: in eukaryotic cells, the cell organelle that is surrounded by two membranes and that is the site of cellular respiration

5. Golgi complex: a cell organelle that helps make and packages materials to be transported out of the cell

6. vesicle: a small cavity or sac that contains materials in a eukaryotic cell

7. lysosome: a cell organelle that contains digestive enzymes

SECTION: THE ORGANIZATION OF LIVING THINGS

1. tissue: a group of similar cells that perform a common function
2. organ: a collection of tissues that carry out a specialized function of the body
3. organ system: a group of organs and tissues that work together to perform body functions
4. organism: a living thing; anything that can carry out life processes independently
5. structure: the arrangement of parts in an organism
6. function: the special, normal, or proper activity of an organ or part

Section Review

SECTION: THE DIVERSITY OF CELLS

1. Sample answer: An organelle is a structure inside a cell that performs a specific function for the cell.
2. Sample answer: Eukaryotic cells have a nucleus, but prokaryotic cells do not.
3. C
4. All organisms are made of one or more cells, the cell is the basic unit of all living things, and all cells come from existing cells.
5. Every cell has a cell membrane, DNA, and cytoplasm.
6. Sample answer: The cell walls and the ribosomes of archaebacteria are different from those structures in eubacteria.
7. Sample answer: The cell is probably a prokaryote; The key is that it does not appear to have a nucleus and its DNA is long and circular.
8. Sample answer: eukaryote: whether the chains were a multicellular organism or a collection of individual cells, membrane-bound organelles, certain types of ribosomes and certain materials in the cell membranes, where the DNA is located, and a nucleus; prokaryote: other types of ribosomes and cell membrane materials, the structure of the DNA, and a nucleus; The first thing I would look for is a nucleus.

9. a typical eubacterial cell; It has no nucleus, and its DNA is long and circular.
10. the flagellum
11. the cell's DNA

SECTION: EUKARYOTIC CELLS

1. Sample answer: Ribosomes are organelles where amino acids are joined together to make proteins. Lysosomes are organelles that carry out cellular digestion. The cell wall is the outermost structure in cells of plants, fungi, and algae.
2. B
3. Sample answer: Golgi complex: packages and distributes proteins within a cell; endoplasmic reticulum: a series of folded membranes on which lipids, proteins, and other materials are made, and through which those materials are delivered to other places in the cell
4. Sample answer: Plant cells have cell walls, but animal cells do not. Plant cells have chloroplasts, which animal cells do not have. Plant cells do not seem to have small lysosomes (they have large vacuoles instead), which animal cells do have.
5. Sample answer: Ribosomes are the organelles where proteins are made. All cells need protein in order to live.
6. Sample answer: Mitochondria are organelles that produce most of a cell's energy. If its mitochondria were destroyed, a cell would eventually die because it would have not been able to produce enough energy to survive.
7. Sample answer: I think plants have an advantage over animals because plants can make their own food just by using sunlight and other nutrients. Animals have to wait for plants to grow in order to get food.
8. This diagram is of an animal cell; the first clue is that the cell has no cell wall.
9. the Golgi complex

SECTION: THE ORGANIZATION OF LIVING THINGS

1. Sample answer: The body has several different kinds of tissue. I think that the most important organ in the body is the brain. Sometimes a part of the body with a certain structure performs more than one function.

2. D

3. $3 \text{ cm} \times 3 \text{ cm} \times 3 \text{ cm} = 27 \text{ cm}^3$
$27 \text{ cm}^3 \div 1 \text{ cm}^3 = 27$ cells; If each side doubles in length, the organism will have 216 cells ($6 \times 6 \times 6 = 216$).

4. Sample answer: Alveoli are tiny sacs whose function is to contain and exchange gases such as oxygen and carbon dioxide. The structure of alveoli, as tiny sacs surrounded by tiny blood vessels, includes the cells that make up the tissue of the alveoli and the tissue that joins the alveoli to the bronchioles, which are part of the lung.The lungs are made of several kinds of tissue, such as the bronchi, bronchioles, and alveoli.

5. Sample answer: The main reason that multicellular organisms can be more complex than unicellular organisms is that multicellular organisms have cell specialization. Specialization allows some cells to do only digestion while others do respiration or circulation. Therefore, the organism is more efficient. Being multicellular also means that an organism may grow larger than a unicellular organism. Size is an advantage because, in general, the larger the organism is, the fewer predators it faces. Finally, being unicellular means that when your one cell dies, you are dead. In a multicellular organism, the death of one cell does not mean the death of the organism.

   Teacher's Note: In fact, only multicellular organisms can have an efficient vascular system, which is the key to efficient delivery of materials to cells and removal of wastes from cells. Most students will probably not know this, but some advanced or interested students may grasp this idea.

Chapter Review

1. cell
2. function
3. organelles
4. eukaryote
5. tissue
6. cell wall
7. C
8. D
9. A
10. B
11. B
12. C
13. Cells must be small to have a large enough surface area-to-volume ratio to get sufficient nutrients and get rid of wastes to survive.
14. Cells are the smallest units of all living things. Cells combine to make tissues. Different tissues combine to make organs, which have specialized jobs in the body. Organs work together in organ systems, which perform body functions.
15. Structure is the shape of a part. Function is the job a part does.
16. The cell membrane separates the cell's contents from the outside environment and the cell membrane controls the flow of nutrients, wastes, and other materials into and out of the cell.
17. The cytoskeleton is a web of tubular and stringy proteins. The cytoskeleton helps give the cell shape and helps the cell move.
18. An answer to this exercise can be found at the end of this book.
19. Answers will vary. Sample answer: The endoplasmic reticulum (ER) is a series of folded membranes within a cell. It is where proteins, lipids, and other materials are made in the cell. The smooth ER also helps break down toxic materials. The ER is the internal delivery system of the cell. The Golgi complex modifies, packages, and distributes proteins to other parts of the cell. It takes materials from the ER and encloses them in a small bubble of membrane. Then, it delivers them to where they are needed in other parts of the cell as well as outside the cell.

20. Answers will vary. Sample answer: The structure of a part is its shape and what it is made of. The function of a part is what that shape and material enable the part to do in the body. For example, alveoli are tiny sacs in the lungs that hold gases. They are made of a membrane that enables oxygen and carbon dioxide to pass in and out of the blood.

21. Answers will vary. Sample answer: Not valid; some organisms are unicellular and have no tissues, organs, or organ systems.

22. Ribosomes make proteins, which all cells and all organisms need to survive. If your ribosomes disappeared, you would die.

23. Answers will vary. Sample answer: Achaebacteria are older because there are many types of methane-making archaebacteria, and because many types of archaebacteria live in very hot places.

24. mitochondrion

25. B

26. C

Reinforcement

BUILDING A EUKARYOTIC CELL

1. B	**7.** B
2. B	**8.** P
3. B	**9.** B
4. B	**10.** B
5. B	**11.** B
6. B	**12.** P

Diagram

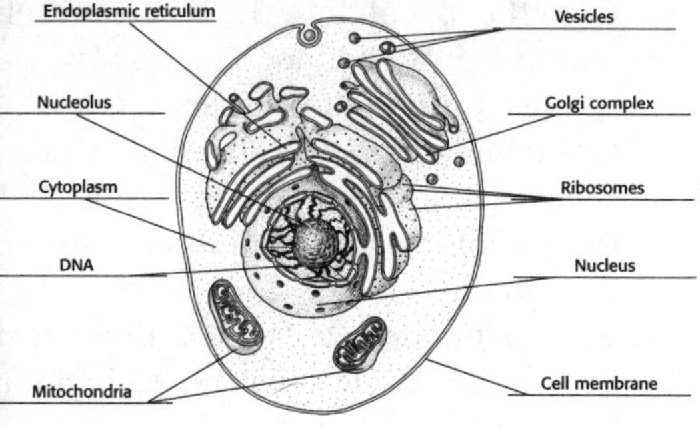

Critical Thinking

1. mitochondria, energy generators— produce energy for work
nucleus, director's office— directs production
lysosome, waste management— collects and destroys wastes
endoplasmic reticulum, materials delivery system—transports materials
Golgi complex, packaging department— processes and packages materials

2. Answers will vary. Sample answer: Both the architectural plans and DNA contain all the information needed for construction. The plans show how materials will flow through a factory. The DNA determines how materials will flow through a cell.

3. Answers will vary. Sample answer: Plant cells have rigid walls for protection. A factory with a more rigid wall around it would be more secure. Plant cells use vacuoles to store water and other materials. It would be useful to have such storage facilities in a factory. Plant cells make food from the sun's energy. A factory that could produce its own solar energy would be more self-sufficient and run more cheaply than a factory without solar energy.

Section Quizzes

SECTION: THE DIVERSITY OF CELLS

1. H	**6.** E
2. J	**7.** G
3. I	**8.** C
4. B	**9.** A
5. F	**10.** D

SECTION: EUKARYOTIC CELLS

1. B	**6.** E
2. A	**7.** H
3. D	**8.** C
4. I	**9.** F
5. J	**10.** G

SECTION: THE ORGANIZATION OF LIVING THINGS

1. C	**6.** A
2. B	**7.** C
3. C	**8.** B
4. C	**9.** D
5. D	**10.** A

Chapter Test A

1. A	**14.** E
2. B	**15.** B
3. D	**16.** A
4. B	**17.** C
5. A	**18.** D
6. C	**19.** B
7. B	**20.** D
8. G	**21.** A
9. F	**22.** C
10. H	**23.** D
11. C	**24.** B
12. A	**25.** C
13. D	

Chapter Test B

1. organs
2. mitochondria
3. nucleus
4. prokaryotic
5. lysosome
6. B
7. D
8. B
9. D
10. C
11. cell membranes, organelles, cytoplasm, DNA
12. cell walls, chloroplasts, chlorophyll
13. Proteins control the chemical reactions in a cell, provide structural support for cells and tissues, and create passageways through the cell membrane.
14. The Golgi complex packages and distributes proteins. It packages proteins in small bubbles made of a piece of membrane. The bubbles then break off and transport their contents to other parts of the cell.
15. cell, tissue, organ, organ system
16. Answers will vary. Sample answer: Prokaryotes are organized only on the cellular level, since they are single-celled organisms. However, some prokaryotic cells do live in colonies. Eukaryotes can be single-celled, but most are multicellular. In more complex eukaryotes, cells form tissues, tissues form organs, and organs work together to form systems.
17. Answers will vary. Sample answer: In a single-celled organism, only one type of cell has to be adapted to the environment. Once that cell has so adapted, it can keep surviving as long as it continues to replicate. In a multicellular organism, there are likely to be many different kinds of cells. To survive in an extreme environment, every kind of cell would have to adapt. This adaptation would also have to happen at the same time, or damage to one type of cell would affect the whole organism.
18. As a cell grows, it produces more waste and requires more nutrients. These materials must be able to pass through the cell membrane. However, as a cell grows, its surface-to-volume ratio decreases. This causes the cell's exchange of materials to become less efficient, which may lead to cell death. For this reason, an embryo divides into many small cells instead of growing into a few very large cells.
19. Alveoli are tiny air sacs that are wrapped in blood vessels. The air sacs can expand as air is drawn into the lungs. Oxygen from the air passes from the alveoli into the blood vessels and then into the blood and body tissues. Carbon dioxide leaves the blood and enters the alveoli as well. When the lungs exhale, the walls of the alveoli collapse and force the carbon dioxide out.
20. **a.** Mitochondria; **b.** DNA; **c.** ribosomes; **d.** proteins; **e.** endoplasmic reticulum; **f.** Golgi complex

Chapter Test C

1. B
2. C
3. C
4. B
5. C
6. A
7. B
8. C
9. A
10. B
11. B
12. D
13. A
14. C
15. A
16. C
17. B
18. cell
19. tissue
20. organ
21. system
22. multicellular
23. structure

Standardized Test Preparation

READING

Passage 1

1. D
2. G
3. B

Passage 2

1. B
2. I
3. B

INTERPRETING GRAPHICS

1. B
2. H
3. A
4. I

MATH

1. B
2. G

Vocabulary Activity

Completed crossword puzzle. Answers:

Across: 3. CYTOPLASM, 7. BACTERIA, 10. NUCLEUS, 11. LYSOSOME, 12. PROKARYOTIC, 13. GOLGI COMPLEX, 17. MULTICELLULAR, 20. CELL, 23. HOOKE, 24. CHLOROPLAST, 25. ORGANISM, 26. ORGAN SYSTEMS, 27. ORGANELLE.

Down: 1. CELL WALL, 2. MITOCHONDRIA, 3. CYTOPLASM, 4. ORGANELLE, 5. RIBOSOMES, 6. TISSUE, 8. EUKARYA, 9. VESICLES, 14. CELL MEMBRANE, 15. NUCLEUS, 16. CELLTYPE, 18. ER, 19. ATP, 21. VACUOLE, 22. DNA.

SciLinks Activity

1. The process of converting oxygen and nutrients to ATP is aerobic respiration.
2. Microtubules are straight, hollow cylinders of protein that are found in the cytoskeleton.
3. Prokaryotes are approximately 2 billion years older than eukaryotes.
4. A typical plant cell has one large vacuole, which stores water and other compounds and helps support the cell.
5. Human DNA is made up of 64 molecules.

Lesson Plan

Section: The Diversity of Cells

Pacing

Regular Schedule: **with lab(s):** 3 days   **without lab(s):** 1 day
Block Schedule: **with lab(s):** 1.5 days   **without lab(s):** 0.5 day

Objectives

1. State the parts of the cell theory.

2. Explain why cells are so small.

3. Describe the parts of a cell.

4. Describe how eubacteria are different from archaebacteria.

5. Explain the difference between prokaryotic cells and eukaryotic cells.

National Science Education Standards Covered

UCP 3: Change, constancy, and measurement

UCP 5: Form and function

SAI 1: Abilities necessary to do scientific inquiry

SAI 2: Understandings about scientific inquiry

ST 2: Understandings about science and technology

SPSP 5: Science and technology in society

HNS 1: Science as a human endeavor

HNS 3: History of science

LS 1a: Living systems at all levels of organization demonstrate the complementary nature of structure and function. Important levels of organization for structure and function include cells, organs, tissues, organ systems, whole organisms, and ecosystems.

LS 1b: All organisms are composed of cells, the fundamental unit of life. Most organisms are single cells; other organisms, including humans, are multicellular.

LS 1c: Cells carry on the many functions needed to sustain life. They grow and divide, thereby producing more cells. This requires that they take in nutrients, which they use to provide energy for the work that cells do and to make the materials that a cell or organism needs.

LS 2c: Every organism requires a set of instructions for specifying traits. Heredity is the passage of these instructions from one generation to another.

LS 3b: Regulation of an organism's internal environment involves sensing the internal environment and changing physiological activities to keep conditions within the range required to survive.

Lesson Plan

LS 5a: Millions of species of animals, plants, and microorganisms are alive today. Although different species might look dissimilar, the unity among organisms becomes apparent from an analysis of internal structures, the similarity of their chemical processes, and the evidence of common ancestry.

KEY

SE = Student Edition   TE = Teacher's Edition
CRF = Chapter Resource File

FOCUS *(5 minutes)*

- **Chapter Starter Transparency** Use this transparency to introduce the chapter.

- **Bellringer, TE** Discuss why cells were not discovered until the 1660s in terms of a technological advance: the invention of the microscope.

- **Bellringer Transparency** Use this transparency as students enter the classroom and find their seats.

MOTIVATE *(10 minutes)*

- **Activity, Modeling Cell Discovery, TE** Before beginning this section, have students model Robert Hooke's discovery. (**GENERAL**)

TEACH *(110 minutes)*

- **Inclusion Strategies, TE** Expand discussion by drawing a timeline and dating the discovery of cells in reference to familiar or key events.

- **Reading Strategy, Prediction Guide, SE** Students predict reasons why most cells are so small.

- **Research, Be a Good Host, TE** Students conduct research and write and present a report on a bacterium. (**GENERAL**)

- **Connection to Language Arts, Smallest Living Thing, TE** Students research the organism considered "the smallest living thing." (**GENERAL**)

- **Activity, Archea, TE** Students determine the similarity of archaebacteria to eubacteria and eukaryotes. (**ADVANCED**)

- **Connection to Earth Science, Discovering Ancient Earth, TE** Student research the work of paleontologists. (**GENERAL**)

- **Chapter Lab, Elephant-Sized Amoebas?, SE** Students explore why a single-celled organism cannot grow to the size of an elephant. (**GENERAL**)

- **Datasheet for Chapter Lab, CRF** Students use the datasheet to complete the chapter lab. (**GENERAL**)

- **Directed Reading A/B, CRF** These worksheets reinforce basic concepts and vocabulary presented in the lesson. (**BASIC/SPECIAL NEEDS**)

- **Vocabulary and Section Summary, CRF** Students write definitions of key terms and read a summary of section content. (**GENERAL**)

CLOSE *(10 minutes)*

_ **Reteaching, Modeling Cells, TE** Ask students to draw labeled pictures or diagrams of typical eukaryotic and prokaryotic cells. **(BASIC)**

_ **Quiz, TE** Students review what Hooke saw and why he interpreted his observations as he did. **(GENERAL)**

_ **Section Review, CRF** Students answer end-of-section vocabulary, key ideas, critical thinking, and interpreting graphics questions. **(GENERAL)**

_ **Section Quiz, CRF** Students answer 10 objective questions about the diversity of cells. **(GENERAL)**

_ **Alternative Assessment, TE** Students write descriptive statements about vocabulary words and guess the words described. **(GENERAL)**

Lesson Plan

Section: Eukaryotic Cells

Pacing

Regular Schedule: **with lab(s):** N/A **without lab(s):** 1 day

Block Schedule: **with lab(s):** N/A **without lab(s):** 0.5 day

Objectives

1. Identify the different parts of a eukaryotic cell.

2. Explain the function of each part of a eukaryotic cell.

National Science Education Standards Covered

UCP 1: Systems, order, and organization

UCP 4: Evolution and equilibrium

UCP 5: Form and function

LS 1b: All organisms are composed of cells—the fundamental unit of life. Most organisms are single cells; other organisms, including humans, are multicellular.

LS 1c: Cells carry on the many functions needed to sustain life. They grow and divide, thereby producing more cells. This requires that they take in nutrients, which they use to provide energy for the work that cells do and to make the materials that a cell or organism needs.

LS 3a: All organisms must be able to obtain and use resources, grow, reproduce, and maintain stable internal conditions while living in a constantly changing external environment.

LS 5a: Millions of species of animals, plants, and microorganisms are alive today. Although different species might look dissimilar, the unity among organisms becomes apparent from an analysis of internal structures, the similarity of their chemical processes, and the evidence of common ancestry.

LS 5b: Biological evolution accounts for the diversity of species developed through gradual processes over many generations. Species acquire many of their unique characteristics through biological adaptation, which involves the selection of naturally occurring variations in populations. Biological adaptations include changes in structures, behaviors, or physiology that enhance survival and reproductive processes in a particular environment.

KEY

SE = Student Edition **TE** = Teacher's Edition

CRF = Chapter Resource File

FOCUS *(5 minutes)*

_ **Bellringer, TE** List, review, and discuss the differences between prokaryotic and eukaryotic cells.

_ **Bellringer Transparency** Use this transparency as students enter the classroom and find their seats.

MOTIVATE *(10 minutes)*

_ **Discussion, TE** Have students discuss how and why they know their own cells are working. **(GENERAL)**

TEACH *(20 minutes)*

_ **Activity, Cellular Sieve, TE** Students observe a strainer to understand how cell membranes act. **(BASIC)**

_ **Inclusion Strategy, TE** Volunteers line to demonstrate the orderly structure of DNA.

_ **Reading Strategy, Prediction Guide, SE** Students predict whether animal cells will be different from plant. **(GENERAL)**

_ **Activity, Making Models, TE** Students make edible cells at home and explain how they have represented each type of organelle. **(GENERAL)**

_ **Directed Reading A/B, CRF** These worksheets reinforce basic concepts and vocabulary presented in the lesson. **(BASIC/SPECIAL NEEDS)**

_ **Vocabulary and Section Summary, CRF** Students write definitions of key terms and read a summary of section content. **(GENERAL)**

_ **Reinforcement, CRF** This worksheet reinforces basic concepts and vocabulary presented in the lesson. **(BASIC)**

_ **SciLinks Activity, Eukaryotic Cells, SciLinks code HSM0541, CRF** Students research Internet sources related to eukaryotic cells. **(GENERAL)**

CLOSE *(10 minutes)*

_ **Alternative Assessment, TE** Students create a cell using a shoebox. **(GENERAL)**

_ **Reteaching, Organelles and Their Functions, TE** Have students create tables showing organelles and their functions. **(BASIC)**

_ **Quiz, TE** Students review the structures and functions of lysosomes, vacuoles, and the cytoskeleton. **(GENERAL)**

_ **Section Review, CRF** Students answer end-of-section vocabulary, key ideas, critical thinking, and interpreting graphics questions. **(GENERAL)**

_ **Section Quiz, CRF** Students answer 10 objective questions about eukaryotic cells. **(GENERAL)**

Lesson Plan

Section: The Organization of Living Things

Pacing

Regular Schedule: **with lab(s):** N/A   **without lab(s):** 1 day

Block Schedule: **with lab(s):** N/A   **without lab(s):** 0.5 day

Objectives

1. List three advantages of being multicellular.

2. Describe the four levels of organization in living things.

3. Explain the relationship between the structure and function of a part of an organism.

National Science Education Standards Covered

UCP 1: Systems, order, and organization

UCP 2: Evidence, models, and explanation

UCP 5: Form and function

LS 1a: Living systems at all levels of organization demonstrate the complementary nature of structure and function. Important levels of organization for structure and function include cells, organs, tissues, organ systems, whole organisms, and ecosystems.

LS 1b: All organisms are composed of cells—the fundamental unit of life. Most organisms are single cells; other organisms, including humans, are multicellular.

LS 1d: Specialized cells perform specialized functions in multicellular organisms. Groups of specialized cells cooperate to form a tissue, such as a muscle. Different tissues are in turn grouped together to form larger functional units, called organs. Each type of cell, tissue, or organ has a distinct structure and set of functions that serves the organism as a whole.

KEY

SE = Student Edition   **TE** = Teacher's Edition

CRF = Chapter Resource File

FOCUS (5 minutes)

Bellringer, TE Discuss the functions of teeth and muscles and how their structures help them perform.

Bellringer Transparency Use this transparency as students enter the classroom and find their seats.

MOTIVATE *(10 minutes)*

_ **Activity, Concept Mapping, TE** Have students draw concept maps showing cell, tissue, organ, and organ system levels of organization. (**GENERAL**)

TEACH *(20 minutes)*

_ **Reading Strategy, Paired Summarizing, SE** Students work in pairs to summarize the material in this section and discuss any ideas that seem confusing to them. (**GENERAL**)

_ **Discussion, Muscles, TE** Have students list and discuss the ways they use their muscles. (**GENERAL/SPECIAL NEEDS**)

_ **Directed Reading A/B, CRF** These worksheets reinforce basic concepts and vocabulary presented in the lesson. (**BASIC/SPECIAL NEEDS**)

_ **Vocabulary and Section Summary, CRF** Students write definitions of key terms and read a summary of section content. (**GENERAL**)

CLOSE *(10 minutes)*

_ **Alternative Assessment, Concept Mapping, TE** Have students make concept map describing functions of organs in an organ system. (**BASIC**)

_ **Reteaching, Levels of Organization, TE** Ask students to list examples of organism parts at all four levels of organization. (**BASIC**)

_ **Quiz, TE** Students review human organ systems and explain differences between unicellular and multicellular organisms. (**GENERAL**)

_ **Section Review, CRF** Students answer end-of-section vocabulary, key ideas, math, and critical thinking questions. (**GENERAL**)

_ **Section Quiz, CRF** Students answer 10 objective questions about the organization of living things. (**GENERAL**)

_ **Homework, Writing, TE** Have students do research to compare structures a fish uses to breathe with those a human uses. (**BASIC**)

Lesson Plan

End of Chapter Review and Assessment

Pacing

Regular Schedule: with lab(s): N/A   **without lab(s):** 2 days

Block Schedule: with lab(s): N/A   **without lab(s):** 1 day

KEY
SE = Student Edition   **TE** = Teacher's Edition
CRF = Chapter Resource File

_ **Standardized Test Preparation, CRF** Students answer reading comprehension, math, and interpreting graphics questions in the format of a standardized test. (**GENERAL**)

_ **Test Generator, One-Stop Planner** Create a customized homework assignment, quiz, or test using the HRW Test Generator program. (**GENERAL**)

_ **Chapter Review, CRF** Students answer end-of-chapter vocabulary, key ideas, critical thinking, and graphics questions. (**GENERAL**)

_ **Vocabulary Activity, CRF** Students review chapter vocabulary terms by completing a puzzle. (**GENERAL**)

_ **Chapter Test A/B/C, CRF** Assign questions from the appropriate test for chapter assessment. (**GENERAL/ADVANCED/SPECIAL NEEDS**)

_ **Performance-Based Assessment, CRF** Assign this activity for general level assessment of the chapter. (**GENERAL**)

_ **CNN Video, CNN Presents Science in the News: Science, Technology & Society,** Segment 3, Flavor Cells; Segment 4, Treating Pets with Laser Light

Cells: The Basic Units of Life

MULTIPLE CHOICE

1. Humans like you are
   a. machines.
   b. systems.
   c. organisms.
   d. protists.
   Answer: C   Difficulty: 1   Section: 3   Objective: 1

2. One benefit of being a large organism is that you have
   a. larger cells.
   b. fewer predators.
   c. simpler functions.
   d. only one kind of cell.
   Answer: B   Difficulty: 1   Section: 3   Objective: 1

3. The life span of a multicellular organism is
   a. only as long as the life of one cell.
   b. shorter than that of a single-celled organism.
   c. not limited to the life of a single cell.
   d. the same in every cell.
   Answer: C   Difficulty: 1   Section: 3   Objective: 1

4. A group of cells with the same function make up
   a. an organism.
   b. an organ system.
   c. a tissue.
   d. a structure.
   Answer: C   Difficulty: 1   Section: 3   Objective: 2

5. In what kind of tissue does photosynthesis take place?
   a. nerve
   b. muscle
   c. transport
   d. ground
   Answer: D   Difficulty: 1   Section: 3   Objective:

6. An organ consists of
   a. two or more tissues.
   b. a group of cells.
   c. two or more systems.
   d. nerves and muscles.
   Answer: A   Difficulty: 1   Section: 3   Objective: 2

7. An organ system has
   a. one kind of tissue.
   b. only one function.
   c. two or more organs.
   d. one main kind of cell.
   Answer: C   Difficulty: 1   Section: 3   Objective: 2

8. Even simple multicellular organisms can have
   a. organs.
   b. specialized cells.
   c. systems.
   d. colonies.
   Answer: B   Difficulty: 1   Section: 3   Objective: 2

9. The highest level of organization is the
   a. cell.
   b. tissue.
   c. organ.
   d. system.
   Answer: D   Difficulty: 1   Section: 3   Objective: 2

10. The functions of an organism's parts are related to those parts'
    a. structures.
    b. systems.
    c. blood cells.
    d. alveoli.
    Answer: A   Difficulty: 1   Section: 3   Objective: 3

11. What is smallest unit that can perform all the processes necessary for life?
    a. cell
    b. nucleus
    c. organelle
    d. protist
    Answer: A   Difficulty: 1   Section: 1   Objective: 1

12. Robert Hooke and Anton van Leeuwenhoek not only helped discover cells but also
    a. discovered that cells came from existing cells.
    b. helped develop the microscope.
    c. concluded that all living things have cells.
    d. discovered mushrooms and fungi.

    Answer: B   Difficulty: 1   Section: 1   Objective: 1

13. Leeuwenhoek called the single-celled organisms that he found in pond scum
animalcules." Today we know them as
    a. animals.   c. fungi.
    b. plant life.   d. protists.

    Answer: D   Difficulty: 1   Section: 1   Objective: 1

14. Scientist Matthias Schleiden contributed to the cell theory by concluding that
    a. the cells of plants and animals were the same.
    b. all plant parts were made of cells.
    c. the cells of plants were different from those of animals.
    d. all animal tissues were made of cells.

    Answer: B   Difficulty: 1   Section: 1   Objective: 1

15. Which of the following statements is not part of the cell theory?
    a. Animals and plants share the same kinds of cells.
    b. All organisms are made up of one or more cells.
    c. The cell is the basic unit of all living things.
    d. All cells come from existing cells.

    Answer: A   Difficulty: 1   Section: 1   Objective: 2

16. Most cells are a very small size because
    a. They don't have hard shells like eggs.
    b. Their volume does not increase.
    c. Their volume is limited by how large their surface area is.
    d. Their surface area-to-volume ratio is too small.

    Answer: C   Difficulty: 1   Section: 1   Objective: 2

17. What cell part supports the cell and might be made of cellulose or chitin?
    a. cell membrane   c. ribosome
    b. cell wall   d. nucleus

    Answer: B   Difficulty: 1   Section: 2   Objective: 2

18. What part of the cell forms a barrier between the cell and its environment?
    a. cell membrane   c. ribosome
    b. cell wall   d. cholesterol

    Answer: A   Difficulty: 1   Section: 2   Objective: 2

19. What part of the cell keeps the cell membrane from collapsing?
    a. cell wall   c. cytoskeleton
    b. cytoplasm   d. nucleus

    Answer: C   Difficulty: 1   Section: 2   Objective: 2

20. A cell's nucleus contains DNA, which carries genetic material with
    a. ribosomes.   c. the endoplasmic reticulum.
    b. the cytoskeleton.   d. instructions for how to make protein.

    Answer: D   Difficulty: 1   Section: 2   Objective: 2

21. Ribosomes, the organelles that make proteins, are found on the membranes of the
    a. cell wall.   c. mitochondria.
    b. endoplasmic reticulum.   d. vacuoles.

    Answer: B   Difficulty: 1   Section: 2   Objective: 2

22. What part of the cell acts as the cell's delivery system?
    a. nucleus
    b. nucleolis
    c. mitochondrion
    d. endoplasmic reticulum

    Answer: D   Difficulty: 1   Section: 2   Objective: 2

23. Energy released by a cell's mitochondrion is stored in
    a. ATP.
    b. DNA.
    c. the ER.
    d. RNA.

    Answer: A   Difficulty: 1   Section: 2   Objective: 2

24. What cell parts carry materials between organelles such as the ER and the Golgi complex?
    a. ribosomes
    b. lysosomes
    c. vesicles
    d. vacuoles

    Answer: C   Difficulty: 1   Section: 2   Objective: 2

25. Larger size, longer life, and specialization are three advantages to being a
    a. eukaryote.
    b. prokaryote.
    c. unicellular organism.
    d. multicellular organism.

    Answer: D   Difficulty: 1   Section: 3   Objective: 1

26. Which of **the** following is true of each of the four levels of organization of living things?
    a. Each contains larger cells than the level below it.
    b. Each is more complex than the level below it.
    c. Each performs the same functions as the level below it.
    d. Each is more specialized than the level below it.

    Answer: B   Difficulty: 1   Section: 3   Objective: 2

27. The function of a part of an organism is related to
    a. its arrangement of cells.
    b. the shape of its parts.
    c. the structure of that part.
    d. its appearance under a microscope.

    Answer: C   Difficulty: 1   Section: 3   Objective: 2

28. Which statement is NOT part of the cell theory?
    a. All organisms are made of one or more cells.
    b. Animal and plant cells contain the same organelles.
    c. The cell is the basic unit of living things.
    d. All cells originate from other cells.

    Answer: B   Difficulty: 1   Section: 1   Objective: 1

29. A cell's volume grows faster than its surface area, so if a cell gets too large
    a. its surface area-to-volume ratio will decrease.
    b. the cell membrane and cell walls will break down.
    c. its outer surface will harden like an eggshell does.
    d. it will not be able to take in enough nutrients or get rid of wastes.

    Answer: D   Difficulty: 1   Section: 1   Objective: 2

30. A large vesicle that aids in digestion within plant cells the way lysosomes do is called
    a. an enzyme.
    b. a vacuole.
    c. a mitochondrion.
    d. a nucleolus.

    Answer: B   Difficulty: 1   Section: 2   Objective: 2

31. Most of a cell's ATP is made and stored in the inner membrane of the
    a. Golgi complex.
    b. nucleus.
    c. endoplasmic reticulum.
    d. mitochondrion.

    Answer: D   Difficulty: 1   Section: 2   Objective: 1

32. Specialization in cells makes tissues, organs, and systems
    a. grow large in size.
    b. produce larger cells.
    c. work more efficiently.
    d. stay healthy.

    Answer: C   Difficulty: 1   Section: 3   Objective: 2

33. Which phrase describes a cell?
    a. is always very small
    b. does everything needed for life
    c. always looks like an egg
    d. is found only in plants

    Answer: B   Difficulty: 1   Section: 1   Objective: 1

34. What are all organisms made of?
    a. plants
    b. protists
    c. cells
    d. eggs

    Answer: C   Difficulty: 1   Section: 1   Objective: 1

35. Where do all cells come from?
    a. animals
    b. ponds
    c. cells
    d. eggs

    Answer: C   Difficulty: 1   Section: 1   Objective: 1

36. What keeps the size of most cells very small?
    a. their hard shells
    b. the surface area–to-volume ratio
    c. food and wastes
    d. their thin surfaces

    Answer: B   Difficulty: 1   Section: 1   Objective: 2

37. What protects the inside of a cell from the outside world?
    a. cytoplasm
    b. nucleus
    c. cell membrane
    d. DNA

    Answer: C   Difficulty: 1   Section: 1   Objective: 3

38. How are archaebacteria different from eubacteria?
    a. Archaebacteria have different ribosomes.
    b. Archaebacteria have only one cell.
    c. Archaebacteria have cell membranes.
    d. Archaebacteria have RNA, not DNA.

    Answer: A   Difficulty: 1   Section: 1   Objective: 4

39. Robert Hooke thought that animals did not have cells because he
    a. had not yet invented the microscope.
    b. could not see animal cells in his microscope.
    c. had not yet discovered protists.
    d. was looking at dead cork cells, not live ones.

    Answer: B   Difficulty: 1   Section: 1   Objective: 1

40. The organisms that Leeuwenhoek called animalcules are today known as
    a. cells.
    b. eukaryotes.
    c. prokaryotes.
    d. protists.

    Answer: D   Difficulty: 1   Section: 1   Objective: 1

41. The cell theory was developed
    a. by Robert Hooke.
    b. by Rudolf Virchow.
    c. over a period of more than 200 years.
    d. in the year 1858.

    Answer: C   Difficulty: 1   Section: 1   Objective: 1

42. Which two things must be compared to explain why almost all cells are small?
    a. surface area and volume
    b. the shell and the yolk
    c. food production and waste elimination
    d. membranes and organelles

    Answer: A   Difficulty: 2   Section: 1   Objective: 2

43. An organelle that is membrane-bound is
    a. part of a prokaryote.
    b. surrounded by membranes.
    c. unable to move around in the cell.
    d. part of the nucleus.
    Answer: B   Difficulty: 2   Section: 1   Objective: 5

44. Both archaebacteria and eubacteria
    a. are commonly known as bacteria.
    b. have ribosomes very much like those of eukaryotes.
    c. have circular DNA.
    d. include species of extremophile organisms.
    Answer: C   Difficulty: 2   Section: 1   Objective: 4

45. Protists are a group of organisms that include
    a. only prokaryotes.
    b. only eukaryotes.
    c. only small organisms found in pond water.
    d. both prokaryotes and eukaryotes.
    Answer: D   Difficulty: 2   Section: 1   Objective: 4

46. The complex sugar cellulose is found in the cell walls of
    a. all prokaryotes.
    b. plants.
    c. animals.
    d. fungi.
    Answer: B   Difficulty: 1   Section: 2   Objective: 2

47. Because lipids are hydrophobic and face inward, their ends
    a. keep water inside the cell.
    b. get rid of wastes.
    c. attract water.
    d. replace cell walls.
    Answer: A   Difficulty: 2   Section: 2   Objective: 2

48. The hydrophilic ends of phospholipics face outward, where they serve to
    a. protect the cell from water.
    b. get rid of wastes.
    c. attract water.
    d. replace cell walls.
    Answer: C   Difficulty: 3   Section: 2   Objective: 2

49. What is cytoplasm?
    a. the nucleus of a cell
    b. the fluid inside a cell
    c. the genetic material in a cell
    d. the proteins in a cell
    Answer: B   Difficulty: 1   Section 1   Objective: 3

50. Where does photosynthesis take place in a cell?
    a. in the nucleus
    b. in the mitochondria
    c. in the chloroplasts
    d. in the ribosomes
    Answer: C   Difficulty: 1   Section: 2   Objective: 2

51. What does the Golgi complex do in a cell?
    a. It packages and distributes proteins.
    b. It is the power source of the cell.
    c. It makes sugar and oxygen.
    d. It makes proteins.
    Answer: A   Difficulty: 1   Section: 2   Objective: 2

52. What is the job of the lysosmes?
    a. They store water.
    b. They digest food particles.
    c. They make new cells.
    d. They package proteins.
    Answer: B   Difficulty: 1   Section: 2   Objective: 2

COMPLETION

Use the terms from the following list to complete the sentences below.

cell membrane	nucleus
reside	ribosome
organelle	Golgi complex
mitochondria	organs
prokaryotic	

53. Various tissues that work together to perform a specific job constitute

_____________________.

Answer: organs   Difficulty: 1   Section: 3   Objective: 2

54. The role of the cell's _____________________ is to release energy that can be used to power various cellular processes.

Answer: mitochondria

Difficulty: 1   Section: 2   Objective: 2

55. DNA, the genetic material in cells, is located in a eukaryotic cell's _____________________.

Answer: nucleus   Difficulty: 1   Section: 2   Objective: 1

56. Cells that have no membrane-covered organelles are _____________________.

Answer: prokaryotic Difficulty: 1   Section: 1   Objective: 5

57. A part of the Golgi complex can pinch off and form a(n) _____________________, which distributes materials to other parts of the cell.

Answer: reside   Difficulty: 1   Section: 2   Objective: 2

Use the terms from the following list to complete the sentences below.

Cell	organ
Structure	tissue
Multicellular	system

58. The lowest level of organization is the _____________________.

Answer: cell   Difficulty: 1   Section: 3   Objective: 2

59. Cells that are like each other and do the same job form a(n) _____________________.

Answer: tissue   Difficulty: 1   Section: 3   Objective: 2

60. A structure made of two or more tissues working together is called a(n)

_____________________.

Answer: organ   Difficulty: 1   Section: 3   Objective: 2

61. A group of organs that work together form an organ _____________________.

Answer: system   Difficulty: 1   Section: 3   Objective: 2

61. Larger size, longer life, and more-specialized cells are characteristics of _____________________ organisms.

Answer: multicellular

Difficulty: 1   Section: 3   Objective: 1

62. How a part of an organism works is related to how it is built, or its

_____________________.

Answer: structure   Difficulty: 1   Section: 3   Objective: 3

SHORT ANSWER

63. What four elements do all cells have in common?
    Answer: cell membranes, organelles, cytoplasm, DNA
    Difficulty: 2   Section: 1   Objective: 3

64. What are three elements that plant cells have and animal cells do not?
    Answer: cell walls, chloroplasts, chlorophyll
    Difficulty: 2   Section: 1   Objective: 3

65. List three roles played by proteins within a cell.
    Answer:
    Proteins control the chemical reactions in a cell, provide structural support for cells and tissues, and create passageways through the cell membrane.
    Difficulty: 2   Section: 2   Objective: 2

66. Identify two functions of the Golgi complex and describe how it performs those functions.
    Answer:
    The Golgi complex packages and distributes proteins. It packages proteins in small bubbles made of a piece of membrane. The bubbles then break off and transport their contents to other parts of the cell.
    Difficulty: 1   Section: 2   Objective: 2

67. List four levels of organization of living things.
    Answer: cell, tissue, organ, organ system
    Difficulty: 1   Section: 3   Objective: 2

68. What is the role of the nucleolis?
    Answer: The nucleolis stores material that will be used to make ribosomes.
    Difficulty: 1   Section: 2   Objective: 2

69. What is an amino acid?
    Answer:
    An amino acid is one of about 20 different organic molecules that are used to make proteins.
    Difficulty: 1   Section: 2   Objective: 2

70. What are the two different kinds of endoplasmic reticulum?
    Answer: rough and smooth
    Difficulty: 1   Section: 2   Objective: 1

71. What are two functions of smooth ER?
    Answer: Smooth ER makes lipids and breaks down toxic materials that enter the cell.
    Difficulty: 1   Section: 2   Objective: 2

72. Besides the nucleus, what are two cell parts that make DNA?
    Answer: mitochondria, chloroplasts
    Difficulty: 2   Section: 2   Objective: 2

73. Name three cell parts that help defend the cell against invading substances.
    Answer: cell membrane, smooth ER, lysosomes
    Difficulty: 2   Section: 2   Objective: 2

74. Name two types of organelles that release energy.
    Answer: mitochondria, chloroplasts
    Difficulty: 2   Section: 2   Objective: 2

75. What kinds of cells enable larger animals like humans to eat a wider variety of prey?
    Answer: specialized cells
    Difficulty: 3   Section: 3   Objective: 1

76. What four basic types of tissues do animals have?
    Answer: nerve, muscle, connective, protective
    Difficulty: 1   Section: 3   Objective: 2

77. What are two elements of a part's structure?
    Answer: the shape of the part and the material from which it is made
    Difficulty: 1   Section: 3   Objective: 3

78. How do your cells compare to the cells of smaller animals in size?
    Answer: The cells of larger and smaller animals are about the same size.
    Difficulty: 2   Section: 3   Objective: 1

79. Why weren't cells discovered until 1665? What invention made their discovery possible?
    Answer:
    Cells weren't discovered until 1665 because almost all cells are too small to be seen
    with the naked eye. The microscope is the invention that made their discovery
    possible.
    Difficulty: 2   Section: 1   Objective: 1

80. When Robert Hooke saw the "juice" in some cells, what was he looking at?
    Answer: cytoplasm
    Difficulty: 2   Section: 1   Objective: 1

81. Why did Hooke think that cells existed only in plants and fungi and not in animals?
    Answer:
    Plant and fungi have cell walls. Hook's microscope wasn't strong enough to view the
    more delicate cell membranes of animal cells.
    Difficulty: 2   Section: 1   Objective: 1

82. List three differences between prokaryotic and eukaryotic cells.
    Answer:
    Prokaryotic cells have circular DNA, no nucleus, and no membrane-covered
    organelles, Eukaryotic cells have linear DNA, a nucleus, and membrane-covered
    organelles.
    Difficulty: 2   Section: 1   Objective: 5

83. Why do scientists sometimes say that plant cell vacuoles are just large lysomes?
    Answer: because vacuoles store digestive enzymes and aid in cellular digestion.
    Difficulty: 2   Section: 2   Objective: 2

84. What is the cytoskeleton?
    Answer:
    a web of proteins in the cytoplasm that gives the cell shape and may help the cell
    move.
    Difficulty: 1   Section: 2   Objective: 1,2

85. Why can't you use your teeth to breathe? Why can't you use your muscles to digest food?
    Answer: because teeth and muscles are not specialized for those functions.
    Difficulty: 2   Section: 3   Objective: 2

86. What is the relationship between your digestive system, stomach, and intestines?

    Answer:
    The digestive system is an organ system. The stomach and intestines are organs within that system.

    Difficulty: 2   Section: 3   Objective: 2

87. What is the main difference between a unicellular organism and a multicellular organism in the way life processes are carried out?

    Answer:
    Sample answer: The main difference is that a unicellular organism must perform all the functions by itself, while a multicellular organism may have specialized cells to carry out different functions.

    Difficulty: 2   Section: 3   Objective: 1

MATCHING

a. archaebacteria	f. organelle
b. cell membrane	g. prokaryote
c. eubacteria	h. ribosomes
d. eukaryote	i. surface area-to-volume ratio
e. nucleus	j. cytoplasm

88. ____ tiny, round organelles made of protein and other material

    Answer: H   Difficulty: 1   Section: 1   Objective: 3

89. ____ the fluid inside a cell

    Answer: J   Difficulty: 1   Section: 1   Objective: 3

90. ____ the reason that most cells are limited to a very small size

    Answer: I   Difficulty: 1   Section: 1   Objective: 2

91. ____ a protective layer that covers the cell's surface and acts as a barrier

    Answer: B   Difficulty: 1   Section: 1   Objective: 3

92. ____ small bodies in a cell's cytoplasm that are specialized to perform specific functions

    Answer: F   Difficulty: 1   Section: 1   Objective: 3

93. ____ in a eukaryotic cell, an organism that contains the cell's DNA and that has a role in growth, metabolism, and reproduction

    Answer: E   Difficulty: 1   Section: 1   Objective: 2

94. ____ an organism that consists of a single cell that does not have a nucleus or membrane-bound organelles

    Answer: G   Difficulty: 1   Section: 1   Objective: 5

96. ____ prokaryotes that are the smallest cells and that have ribosomes

    Answer: C   Difficulty: 1   Section: 1   Objective: 4

97. ____ prokaryotes that include extremophiles, organisms that live in extreme conditions

    Answer: A   Difficulty: 1   Section: 1   Objective: 4

98. ____ an organism made up of cells that have a nucleus enclosed by a membrane as well as membrane-bound organelles

    Answer: D   Difficulty: 1   Section: 1   Objective: 5

a. cell membrane	f. Golgi complex
b. cell wall	g. lysosomes
c. chloroplasts	h. mitochondrion
d. cytoskeleton	i. nucleus
e. endoplasmic reticulum	j. ribosomes

99. _____ a rigid structure that gives support to a cell

    Answer: B   Difficulty: 1   Section: 2   Objective: 2

100. _____ a barrier that encloses and protects the cell

    Answer: A   Difficulty: 1   Section: 2   Objective: 2

101. _____ a web of proteins in the cytoplasm that keeps a cell's membrane from collapsing

    Answer: D   Difficulty: 1   Section: 2   Objective: 2

102. _____ a large organelle that produces and stores the cell's DNA

    Answer: I   Difficulty: 1   Section: 2   Objective: 2

103. _____ organelles that make proteins

    Answer: J   Difficulty: 1   Section: 2   Objective: 2

104. _____ a system of folded membranes that functions as the internal delivery system of a cell

    Answer: E   Difficulty: 1   Section: 2   Objective: 2

105. _____ an organelle that functions as the main power source of a cell, breaking down sugar to produce energy

    Answer: H   Difficulty: 1   Section: 2   Objective: 2

106. _____ organelles in which photosynthesis takes place

    Answer: C   Difficulty: 1   Section: 2   Objective: 2

107. _____ the organelle that packages and distributes proteins

    Answer: F   Difficulty: 1   Section: 2   Objective: 2

108. _____ organelles that contain digestive enzymes

    Answer: G   Difficulty: 1   Section: 2   Objective: 2

a. archaebacteria   e. multicellular
b. cell membrane   f. nucleus
c. eubacteria   g. organelles
d. eukaryotes   h. prokaryote

109. _____ the part of the cell that keeps the cytoplasm inside and controls materials going in and out of the cell

    Answer: B   Difficulty: 1   Section: 1   Objective: 3

110. _____ structures that are usually surrounded by membranes and which perform specific functions within the cell

    Answer: G   Difficulty: 1   Section: 1   Objective: 3

111. _____ an organelle that contains the cell's deoxyribonucleic acid

    Answer: F   Difficulty: 1   Section: 1   Objective: 3

112. _____ a single-celled organism that has no nucleus or membrane-bound organelles

    Answer: H   Difficulty: 1   Section: 1   Objective: 5

113. _____ the smallest and most common form of prokaryotes, containing DNA, ribosomes, and a flagellum

    Answer: C   Difficulty: 1   Section: 1   Objective: 4

114. _____ prokaryotes that include such types as heat-loving, salt-loving, and methane-making

    Answer: A   Difficulty: 1   Section: 1   Objective: 4

115. _____ organisms made up of cells that have a nucleus and membrane-bound organelles

    Answer: D   Difficulty: 1   Section: 1   Objective: 5

116. _____ word that describes most organisms that you can see with your naked eye

    Answer: E   Difficulty: 1   Section: 1   Objective: 5

a. DNA   c. nucleus
b. eukaryote   d. prokaryote

117. _____ a cell with a nucleus

    Answer: B   Difficulty: 1   Section: 1   Objective: 5

118. _____ a cell without a nucleus

    Answer: D   Difficulty: 1   Section: 1   Objective: 5

119. ____ genetic material in cells
     Answer: A   Difficulty: 1   Section: 1   Objective: 3
120. ____ where DNA is stored
     Answer: C   Difficulty: 1   Section: 1   Objective: 3

     a. cell walls   c. ribosome
     b. endoplastmic reticulum

121. ____ stiff surfaces that supportcells
     Answer: A   Difficulty: 1   Section: 2   Objective: 2
122. ____ organelle that makes proteins
     Answer: C   Difficulty: 1   Section: 2   Objective: 2
123. ____ a cell's delivery system
     Answer: B   Difficulty: 1   Section: 2   Objective: 2

     a. found in plant cells   c. found in both plant and animal cells
     b. found in animal cells

124. ____ cellulose, a complex sugar used to build cell walls
     Answer: A   Difficulty: 2   Section: 2   Objective: 2
125. ____ nucleus
     Answer: C   Difficulty: 2   Section: 2   Objective: 2
126. ____ organelles that aid photosynthesis
     Answer: A   Difficulty: 1   Section: 2   Objective: 2
127. ____ vacuoles, large vesicles that store water and other materials
     Answer: C   Difficulty: 1   Section: 2   Objective: 2

ESSAY

28. Compare the levels of organization among eukaryotes with the types of organization found among prokaryotes.

Answer:

Answers will vary. Sample answer: Prokaryotes are organized only on the cellular level, since they are single-celled organisms. However, some prokaryotic cells do live in colonies. Eukaryotes can be single-celled, but most are multicellular. In more complex eukaryotes, cells form tissues, tissues form organs, and organs work together to form systems.

Difficulty: 3   Section: 3   Objective: 2

29. Why does being single-celled give a species of extremophiles a greater chance of survival? To answer the question, review your answer to question 16 and imagine a multicellular species in the same environment.

Answer:

Answers will vary. Sample answer: In a single-celled organism, only one type of cell has to be adapted to the environment. Once that cell has so adapted, it can keep surviving as long as it continues to replicate. In a multicellular organism, there are likely to be many different kinds of cells. To survive in an extreme environment, every kind of cell would have to adapt. This adaptation would also have to happen at the same time, or damage to one type of cell would affect the whole organism.

Difficulty: 3   Section: 1, 3   Objective: 2

30. Explain why the cells in an embryo will grow no larger than a certain size before they divide.

Answer:

As a cell grows, it produces more waste and requires more nutrients. These materials must be able to pass through the cell membrane. However, as a cell grows, its surface-to-volume ratio decreases. This causes the cell's exchange of materials to become less efficient, which may lead to cell death. For this reason, an embryo divides into many small cells instead of growing into a few very large cells.

Difficulty: 2   Section: 1   Objective: 2

31. Explain how the structure of alveoli makes their function possible.

Answer:

Alveoli are tiny air sacs that are wrapped in blood vessels. The air sacs can expand as air is drawn into the lungs. Oxygen from the air passes from the alveoli into the blood vessels and then into the blood and body tissues. Carbon dioxide leaves the blood and enters the alveoli as well. When the lungs exhale, the walls of the alveoli collapse and force the carbon dioxide out.

Difficulty: 1   Section: 3   Objective: 3

CONCEPT MAPPING

132. Use the following terms to complete the concept map below:

DNA
Mitochondria
ribosomes

Golgi complex
proteins

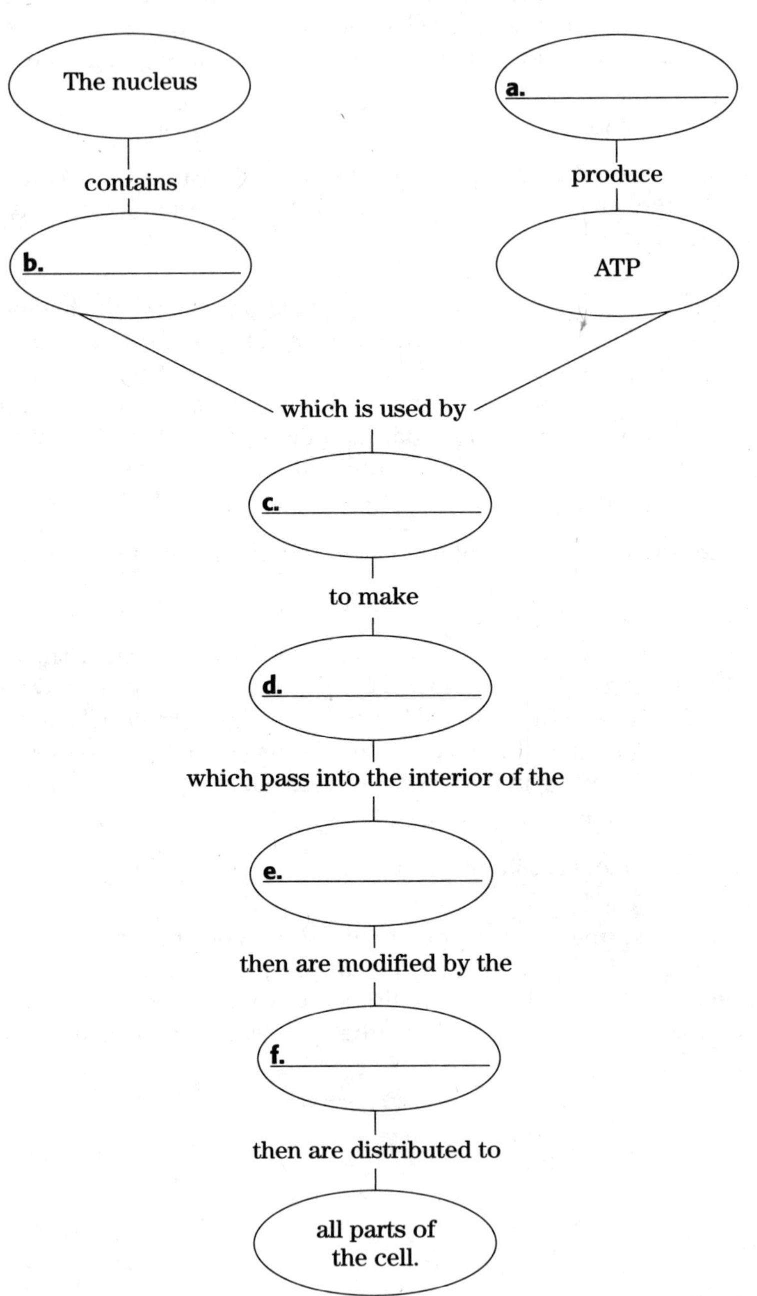

Answer:
Mitochondria; b. DNA; c. ribosomes; d. proteins; e. endoplasmic reticulum; f. Golgi complex

Difficulty: 2   Section: 2   Objective: 2

9997281705